Lecture Notes in Mathematics

Volume 2377

This series reports on new developments in all areas of mathematics and their applications - quickly, informally and at a high level. Mathematical texts analysing new developments in modelling and numerical simulation are welcome. The type of material considered for publication includes:

1. Research monographs
2. Lectures on a new field or presentations of a new angle in a classical field
3. Summer schools and intensive courses on topics of current research.

Texts which are out of print but still in demand may also be considered if they fall within these categories. The timeliness of a manuscript is sometimes more important than its form, which may be preliminary or tentative. Please visit the LNM Editorial Policy (https://drive.google.com/file/d/1MOg4TbwOSokRnFJ3ZR3ciEeKs9hOnNX_/view?usp=sharing)

Titles from this series are indexed by Scopus, Web of Science, Mathematical Reviews, and zbMATH.

Maxime Zavidovique

Discrete Weak KAM Theory

An Introduction through Examples and its Applications to Twist Maps

Springer

Maxime Zavidovique
Institut de Mathématiques de Jussieu
Sorbonne Université
Paris, France

ISSN 0075-8434 ISSN 1617-9692 (electronic)
Lecture Notes in Mathematics
ISBN 978-3-031-96808-2 ISBN 978-3-031-96809-9 (eBook)
https://doi.org/10.1007/978-3-031-96809-9

Mathematics Subject Classification: 49-xx, 37J51, 49L12, 35F21, 35D40, 37E45

This book includes a Foreword written by Albert Fathi.

This Springer imprint is published by the registered company Springer Nature Switzerland AG
The registered company address is: Gewerbestrasse 11, 6330 Cham, Switzerland

To my loving parents for their infallible support

Foreword

What you hold in your hand is a quite valuable addition to the mathematical literature.

Weak KAM theory is the connection between the (Aubry-)Mather theory of Lagrangian Systems and the Crandall-Lions theory of viscosity solutions of the Hamilton-Jacobi equation.

Anybody who tried to teach weak KAM theory faced a big challenge: rarely (at the graduate level at least) people in the audience know of both the Calculus of Variations of paths (up to Tonelli work on existence), rudiments of Dynamical Systems, and the theory of viscosity solutions of the Hamilton-Jacobi equation. Of course, most people have been exposed to the elementary Calculus of Variations that is taught at the undergraduate level. They also know some PDE, but viscosity solutions are rarely considered: In fact, a referee wrote once that from the PDE point of view this theory is not useful since it applies to only one equation. Today at the graduate level most of students have had some exposure to the rudiments of Dynamical Systems.

Therefore the lecturer is faced with the fact that he must cover both Calculus of Variations and the theory viscosity solutions of the Hamilton-Jacobi equation. Usually the material takes most of the semester leaving not much time to develop the weak KAM theory connecting both subjects.

In fact, weak KAM theory has a discrete version that requires almost no prerequisites beyond the common knowledge of Topology and Real Analysis. Maxime Zavidovique is one of the main actors of this discrete version. Hence he very cleverly decided to cover this aspect of weak KAM theory, in which you in fact use all the main tools and ideas of the subject. Therefore after this set of lectures, the reader can more easily understand the (non-discrete) weak KAM theory after covering some of the prerequisites.

There is more in this set of lectures. First carefully crafted and detailed examples, so necessary to understand a general theory. Then a deep study of twist maps. Finally Maxime compares the non-discrete and discrete weak KAM theory, introducing the reader to the non-discrete theory, stressing also some of the more recent developments that he could not cover in detail.

I am sure that these notes will definitely be one of, if not, the main introduction to weak KAM theory. With a bit of envy, I wish I could have done such a good job!

Paris, France
April 2025

Albert Fathi

Remerciements

Je voudrais avant tout remercier Thétis. C'est probablement grâce à sa négligence que j'ai eu le temps de me remettre à la rédaction de ce mémoire, confortablement immobilisé dans mon canapé.

As far as real living persons are concerned, my gratitude goes to Alessio Figalli, Hitoshi Ishii and Claude Viterbo for sacrificing part of their summer time to the task of reading my prose, thank you.

Je suis à jamais redevable à Albert Fathi de m'avoir introduit dans le monde KAM faible, il y a maintenant quinze ans. C'est un monde d'idées simples et profondes, un monde parfois technique mais rarement trop. L'explorer est une jubilation.

Je suis également infiniment reconnaissant à Marie-Claude Arnaud pour les réflexions et résultats partagés. L'énergie communicative avec laquelle tu t'attaques à nos problèmes est motrice et me transporte.

Questo testo è nato dopo un mini corso dato a Cortona en 2015. Inoltre di mangiare tartufo con Montepulciano, mi ha fatto pensare a come organizzare la teoria di maniera pedagogica. Grazie agli organizzatori Andrea, Antonio e Alfonso per questa stupende conferenza.

Ce texte, et une partie de son contenu, n'aurait pas été écrit sans mes coauteurs, en particulier, Andrea, Albert, Antonio et bien sûr Marie-Claude. Apprendre grâce à vous et découvrir ensemble de nouveaux théorèmes suscite chez moi émerveillement et plaisir. Les mathématiques se dégustent en groupe !

Parlant d'équipe, merci au très regretté Nicolas Bergeron, à Alexandru Oancea et Patrice Le Calvez pour votre accompagnement et soutien à Jussieu. Alex, merci aussi et surtout pour tes précieux conseils et la relecture de mon texte. Évoluer dans un environnement professionnel chaleureux et apaisé est un luxe[1] dont j'ai la chance de bénéficier. C'est grâce à tous mes collègues de AA. Une spéciale dédicace à Antonin et Pierre, merci pour votre aide précieuse avec les figures ! Je n'oublie pas

[1] que SU facture 380€.

les anciens collègues, Hélène, Erwan et surtout Grégory pour la formation continue en chocolat et pâtisserie, et pour tout le reste.

Enfin merci à mes parents pour le soutien éternel et la relecture pointilleuse du manuscrit.

Paris, France
April 2025

Maxime Zavidovique

Declarations

Competing Interests The author has no competing interests to declare that are relevant to the content of this manuscript.

Contents

Chapter 1
Introduction

Abstract The aim of these notes is to present a self contained account of discrete weak KAM theory. Put aside the intrinsic elegance of this theory, it is also a toy model for classical weak KAM theory, where many technical difficulties disappear, but where central ideas and results persist. It can therefore serve as a good introduction to (continuous) weak KAM theory. After a general exposition of the general abstract theory, several examples are studied. The last section is devoted to the historical problem of conservative twist maps of the annulus. At the end of the first three Chapters, the relations between the results proved in the discrete setting and the analogous theorems of classical weak KAM theory are discussed. Some key differences are also highlighted between the discrete and classical theory. An appendix briefly comes back to the etymology of weak KAM theory and non exhaustively presents some of its recent striking developments.

The present text initially emerged from lecture notes of a course given in Cortona in 2015 at the INDAM meeting entitled The Hamilton–Jacobi Equation: At the crossroads of PDE, dynamical systems and geometry. The goal of the lectures was to give a complete and thorough introduction to weak KAM theory through the prism of its discrete pendant. It culminated with the proof of convergence of the solutions of the discounted equation, which was new at the time.

The pedagogical motivation is that discrete weak KAM theory is peculiarly elementary. Basic topology is the only prerequisite and the most advanced tools are the Arzelà–Ascoli Theorem and weak compactness of probability measures on a compact metric space. However, all important results of weak KAM theory find their analogue in discrete weak KAM theory and the proofs being stripped of technicalities reveal the key ideas more clearly. After studying this toy model, the interested reader can then go on to learn more on the major theories, as Calculus of Variations, Viscosity solutions of Hamilton–Jacobi equations, and Tonelli Hamiltonian Dynamical Systems. The latter are at the core of classical weak KAM theory.

M. Zavidovique, *Discrete Weak KAM Theory*, Lecture Notes in Mathematics 2377,
https://doi.org/10.1007/978-3-031-96809-9_1

We wish to start by explaining in which context weak KAM theory emerged in the 90's. Most notions described will be rigorously defined later in the body of the text.

1.1 Weak KAM, a Bridge Between Aubry–Mather and Viscosity Solutions

Originally discovered by Albert Fathi in the 90's [104–107], weak KAM theory was designed to understand the dynamical objects of Aubry–Mather Theory for Tonelli Hamiltonian/Lagrangian systems through particular functions called *critical subsolutions* and *weak KAM solutions*. As explained by Fathi, the term weak KAM was chosen as KAM tori give rise to strong solutions of the Hamilton–Jacobi equation while weak KAM solutions are weak solutions of the same equation. It turned out that weak KAM solutions and subsolutions fall in the realm of Viscosity Solutions and subsolutions of Hamilton–Jacobi equations, a theory founded by Crandall and Lions [86].

1.1.1 Aubry–Mather Theory

This theory originated in the study of conservative twist maps and Frenkel–Kontorova models. The objective was to understand invariant sets and the structure of minimizing orbits of such systems. Another type of problem was to construct diffusion orbits connecting invariant sets, or to understand obstructions for such orbits to exist. Aubry and Le Daeron [27] and Mather [159] understood that orbits and invariant sets minimizing a certain energy verify similar properties as orbits of homeomorphisms of the circle. Hence they could apply Poincaré-Denjoy theory to classify and understand such invariant sets according to their rotation number. They went on to study the minimal average action of such minimizing orbits (a function only depending on the rotation number), invariant minimal measures (that would become Mather measures) and Mather gave several definitive answers to questions on the existence of connecting orbits [160, 162–164, 166]. Amongst other important contributions let us mention Bangert [30, 31].

Motivated by Moser, Mather developed a generalization of his theory to higher dimensional settings by introducing Minimizing Measures for a Tonelli Lagrangian defined on the tangent bundle of a compact manifold [165]. His next goal would then be to use this tool to tackle Arnol'd diffusion of such systems, a phenomenon highlighted by Arnol'd in his famous examples [26]. Let us present the philosophy. In an integrable system, all orbits are bounded and periodic or quasi-periodic. If one perturbs such a system, KAM theory implies that many quasi-periodic orbits persist. However, Arnol'd constructed examples where for small perturbations, some orbits have a huge drift in energy. He then conjectured that such a phenomenon should be typical. Mather led the way proposing groundbreaking variational mechanisms to construct generic diffusion [167, 169, 171]. Since then, there has been a huge literature trying to carry on Mather's program for diffusion.

1.1.2 Viscosity Solutions

As already mentioned, viscosity solutions were introduced by Crandall and Lions. They provide a simple, robust definition for weak solutions of first and second order PDE's. Strong existence and uniqueness results are obtained for wide classes of equations, including Hamilton–Jacobi equations (stationary and evolutionary) making the solutions worthwhile studying. Let us mention amongst many others the founding works of Ishii, Crandall, Lions [87, 133, 134, 139]. The definition, that makes use of test functions that are super-tangent or sub-tangent to the solution is very geometric, and allows easily to obtain stability results for viscosity solutions. References to learn more about basic (and more advanced) properties are [33, 88, 151]. For example, viscosity solution theory is so flexible as to apply to non-continuous functions. Stability allows to obtain very general convergence theorems of approximation schemes [34, 192]. Other references making use of this idea, in more weak KAM or variational contexts, are [60, 186, 200, 204].

Very soon in the development of the theory of Viscosity Solutions, Lions and collaborators realized the important role of dynamical programming properties for evolutionary equations and the links with Optimal Control theory that naturally emerge. Indeed, the Value function in optimal control is almost systematically a viscosity solution to some Hamilton–Jacobi equation [32]. This is fundamental in weak KAM theory as the Lax–Oleinik semigroup rediscovered by Fathi turns out to be the value function of an optimal control problem.

Another important problem that was solved early on by viscosity solution methods is that of Homogenization of the Hamilton–Jacobi equation. Consider a continuous, $\mathbb{Z}^N$-periodic in the first variable Hamiltonian $H : \mathbb{R}^N \times \mathbb{R}^N$ that is uniformly coercive in the second variable. Fix a bounded and uniformly continuous initial data $u_0 : \mathbb{R}^N \to \mathbb{R}$. Given $\varepsilon > 0$, Lions, Papanicolaou and Varadhan consider the evolutionary Hamilton–Jacobi equation

$$\begin{cases} \partial_t U + H(\frac{x}{\varepsilon}, \partial_x U) = 0, \\ U(0, \cdot) = u_0, \end{cases} \tag{EHJε}$$

that admits a unique viscosity solution $U_\varepsilon : [0, +\infty) \times \mathbb{R}^N \to \mathbb{R}$. They prove that as $\varepsilon \to 0$, the functions U_ε converge locally uniformly to a function U_0 that is characterized as the unique solution to

$$\begin{cases} \partial_t U + \overline{H}(\partial_x U) = 0, \\ U(0, \cdot) = u_0, \end{cases} \tag{E$\overline{H}$J}$$

where $\overline{H}$ is called the *effective Hamiltonian*. It is defined as follows: for $P \in \mathbb{R}^N$, $\overline{H}(P)$ is the only constant such that the cell problem $H(x, P + \partial_x u) = \overline{H}(P)$ admits a $\mathbb{Z}^N$-periodic viscosity solution $u : \mathbb{R}^N \to \mathbb{R}$. In Homogenization theory, the state variable $x \in \mathbb{R}^N$ takes values in the universal cover of the flat torus $\mathbb{T}^N$. A fundamental domain of the covering map is then the cell $[0, 1)^N$ in the sense that the knowledge of a $\mathbb{Z}^N$-periodic function on $\mathbb{R}^N$ is equivalent to its restriction to the cell $[0, 1)^N$. The name cell problem comes from the fact that the unknown function is defined on a cell.

This effective Hamiltonian (in the case of convex Hamiltonians) coincides with the minimal average action, Mather's α function. The solutions to the cell problem in our terminology will be weak KAM solutions. This problem was revisited by Evans in [101] where he introduced the perturbed test function method. Much later, it was studied under the light of symplectic topology by Viterbo [194] and Montzner et al. [176]. Amongst many other follow ups in the spirit of Lions Papanicolaou and Varadhan, let us mention recent generalizations to other manifolds [85, 191].

1.1.3 The Bridge

As all good bridges, weak KAM theory quickly helped the development of both banks it joins. From a dynamical point of view, Fathi's first achievement was to construct connecting orbits through conjugate pairs of weak KAM solutions [104, 106].

From the PDE side, he proved long time convergence of solutions to the evolutionary Hamilton–Jacobi equation on compact manifolds [107], for autonomous Hamiltonians. Though partial results had been obtained by PDE means [179], the new idea he imported from the dynamical world is that long minimizing trajectories tend to accumulate on the support of minimizing Mather measures. This was followed by many generalizations, for instance [35, 89, 135] where variational and PDE methods allow to weaken regularity hypotheses that Dynamical methods require. In a sense, this culminates in works (the first one being [62]) which use an idea of Evans [102] where Mather measures are given a PDE definition and henceforth adapted to more general settings. Other related results are:

- for counterexamples in the non-autonomous setting [36, 111];
- for positive results in dimension 1 [41, 51].

Another question raised in the theory of Hamilton–Jacobi equations was that of regularity of critical subsolutions. For Tonelli Hamiltonians, this was settled by Fathi and Siconolfi in [114] where existence of C^1 subsolutions is proved. Along the way, the authors also gave a simple proof of Mañé's characterizations of Mather measures, as closed minimizing measures [153, 154]. This is the point of view used in this work. In [115] the same authors extended their results to Lipschitz Hamiltonians. Finally, in [42], existence of $C^{1,1}$ subsolutions is obtained by some

Lasry–Lions type approximation [45, 149]. This could be expected as Fathi had also proved that C^1 solutions are automatically $C^{1,1}$ [108].

The latest breakthrough of weak KAM theory in the PDE theory of Hamilton–Jacobi equations is probably the proof of convergence of the solutions of the discounted equations [94], following some special cases in [143]. The discounting method was an approximation procedure used already in [152] to prove existence of solutions to the cell problem. It has the advantage to approximate it by equations verifying a strong comparison principle and having exactly one solution. The convergence result of Davini, Fathi, Iturriaga and the author is that, as the perturbation goes to 0, a particular weak KAM solution is selected when the Hamiltonian is convex in the second variable. Prior conditions on selected limiting solutions had been found by Gomes in [127]. The result strikes by its generality (little regularity is assumed on the Hamiltonian, no strict convexity) and by the flexibility of its proof. It was naturally followed by numerous generalizations and adaptations. Among them let us mention:

- the discrete setting that is presented later in this text [94],
- adaptations to Neumann problems [1],
- non-compact settings [140],
- more abstract duality methods [141, 142],
- second order Hamilton–Jacobi equations [173],
- weakly coupled systems of Hamilton–Jacobi equations [93] using weak KAM tools from [96] and [136, 138] for more general results,
- convergence from the negative direction [90],
- for more general nonlinear discount approximations [76–78, 199],
- for discounted approximations on networks instead of manifolds [183],
- for mean field games [73]
- more recently, with degenerate discounting approximations [205].

Some limitations and counterexamples exist nevertheless when convexity is dropped [137, 208] and last but not least, let us quote [19] for a more geometric perspective on the convergence result.

Crossing back to the dynamical systems world, Patrick Bernard used weak KAM solutions to push Mather's ideas in [41, 43]. Probably the most definitive works on Arnol'd diffusion to this day are [44, 55]. Those works mix on the one hand dynamical strategies dating back to Arnol'd following chains of normally hyperbolic objects having transverse intersections of stable and unstable manifolds, and on the other hand analytical tools about regularity of weak KAM solutions and subsolutions.

In his founding works [153, 154], Mañé proved that a generic Hamiltonian (in a sense he defined), has a unique Mather measure that is hence ergodic for the Lagrangian flow. He then asked if this measure is generically concentrated on a hyperbolic periodic orbit. This question is known as Mañé's conjecture. It is still open but important steps were made by Figalli and Rifford [121, 122] who managed to bring together methods of optimal control and PDE with the dynamical system theory of orbit closing lemma. They prove the conjecture in low regularity assuming the existence of a smooth critical subsolution. Then with Contreras [84] they brought in Arnaud's theory of Green bundles [5] to prove that generically, the Aubry set is hyperbolic if the underlying manifold has dimension 2. The idea is that in this setting, if the Aubry set is not hyperbolic, then positive and negative Green bundles coincide and weak KAM solutions gain some extra regularity allowing to use ideas from the previous works.

Mather had also raised a similar question about obtaining a uniform bound on the number of ergodic Mather measures when the cohomology varies, for generic Hamiltonians. This was solved by Bernard and Contreras [46, 50]. More generally, understanding the shape of the Aubry set is a challenging question that is still to be understood. Progress on the quotiented Aubry set was obtained by Sorrentino and Fathi, Figalli, Rifford in [118, 189] and for the actual Aubry set, by Arnaud [13].

1.2 Weak KAM, Beyond Hamilton–Jacobi Equations

It turns out that the philosophy of weak KAM theory and of Aubry–Mather theory applies to a variety of other areas.

1.2.1 Optimal Control Theory

Weak KAM theory's starting point is the fact that solutions to Hamilton–Jacobi equations of the form (EHJε) (seen in the context of homogenization), on a manifold M, when the Hamiltonian has some convexity properties, are given by an explicit formula called *Lax–Oleinik semigroup* (for $\varepsilon = 1$):

$$U(t,x) = \inf_{\substack{\gamma:[-t,0]\to M\\ \gamma(0)=x}} u_0\big(\gamma(-t)\big) + \int_{-t}^{0} L\big(\gamma(s),\dot\gamma(s)\big)\mathrm{d}s, \tag{1.1}$$

where the infimum is taken amongst all absolutely continuous curves. This formula (or its variations) is a convolution in the (min,+) semiring and is widely studied in Optimal Control theory. In this setting, U is called the *value function*. Excellent introductory references on the subject are [32, 67, 81] and we already mentioned how Fathi and others made groundbreaking progress by importing ideas from Optimal Control (such as semiconcavity, regularity properties of minimizers ...). Contributions in the other directions are also worth highlighting. Recently for instance, conjectures about propagation of singularities of solutions of Hamilton–Jacobi equations were solved by the use of the positive Lax–Oleinik semigroup [65, 70]. Such results find amazing consequences in Riemannian geometry when applied to singularities of distance functions [72].

1.2.2 Contact Type and Systems of Hamilton–Jacobi Equations

Generalizations of weak KAM theory concern more general types of equations. The first family of generalizations we have in mind is that of contact type equations. This means that the Hamiltonian function $H : T^*M \times \mathbb{R} \to \mathbb{R}$ also depends on the value $u(t, x)$ of the unknown function. The terminology comes from the equations of characteristics that preserve a contact form (instead of the symplectic form for classical Hamiltonian equations). In this context, solutions are given by an implicit Lax–Oleinik semigroup and weak KAM arguments can therefore be used. This was exploited in many recent works including [71, 77, 78, 196–198, 207].

In a maybe more surprising way, weak KAM ideas also apply to some weakly coupled systems of Hamilton–Jacobi equations. This is more unexpected as weak KAM makes strong use of the order structure of $\mathbb{R}$ that is less clearly adaptable for systems where the values taken are in $\mathbb{R}^d$ for some $d > 1$. The first evidence of such a link was present in [63, 64] and further developed in [91, 96, 131, 174, 188].

1.2.3 Lorentzian Geometry and Lyapunov Functions

As can be observed already from the definition of the Lax–Oleinik semigroup (1.1) weak KAM theory studies minimization problems, objects verifying a family of inequalities and minimizers of those inequalities. As a matter of fact, central objects in weak KAM theory are critical subsolutions. They are functions $u_0 : M \to \mathbb{R}$ for which the associated function $U(t, x) - \alpha(0)t$ (where U is given by the Lax–Oleinik semigroup (1.1) and $\alpha(0)$ is called the *critical constant*) is non-decreasing in t. Such functions are also characterized by the following two properties:

- u_0 is Lipschitz continuous,
- $(x, D_x u_0) \in H^{-1}(-\infty, \alpha(0)] \cap \{x\} \times T_x^* M$ for almost every $x \in M$.

When H is convex in the second variable, the sets $H^{-1}(-\infty, \alpha(0)] \cap \{x\} \times T_x^* M$ are convex for all x. Therefore a natural generalization is to replace the Hamiltonian by a family of convex sets $C_x \in T_x^* M$ that verify suitable regularity properties. The question being to know if solutions exist, one studies differentiable inclusions of the form

- u is Lipschitz continuous,
- $D_x u \in C_x$ for almost every $x \in M$.

In such settings, objects from Aubry–Mather theory such as the Aubry set appear as obstructions to finding such functions that verify extra conditions (as smoothness or replacing C_x by its interior). Moreover, when possible, weak KAM methods make it possible to construct C^1 solutions to such differentiable inclusions. This was noticed by Fathi and Siconolfi [116] who applied this philosophy to Lorentzian geometry, where the sets C_x are cones provided by a section of non positive 2-forms. This line of research was since developed in [52, 53, 193].

Another fruitful extension of weak KAM ideas concerns Lyapunov functions and is closer to our subject of discrete weak KAM theory. Indeed, given a continuous transformation F of a metric space X, a Lyapunov function is $f : X \to \mathbb{R}$ that is non-increasing (or when possible decreasing) on the orbits of F. That is f verifies the family of inequalities

$$\forall x \in X, \quad f \circ F(x) \leqslant f(x).$$

Clearly, no function can be decreasing on a periodic orbit. More generally any reasonable notion of recurrence will provide an obstruction to the existence of strict Lyapunov functions. Hence being able to construct optimal Lyapunov functions is an important challenge related to fine dynamical properties. The study of this problem, importing weak KAM ideas, was done in Pageault's PhD that led to recovering earlier results of Conley, Akin, Auslander ... in simpler and more precise form. Results are to be found in [112, 113, 181] and were followed by further studies [56–58].

Let us also mention links with ergodic optimization and the analogue of Mañé's conjecture that was recently solved by Contreras [82]. Other works related to weak KAM theory are [123, 124].

1.2.4 Optimal Transportation

Let us describe the original Monge problem in optimal transportation [175]. The goal is to move some material described by a probability measure μ on a space X into a certain configuration described by another probability measure ν on a space Y, by means of a transport map $T : X \to Y$, knowing that the cost to move material from $x \in X$ to $y \in Y$ is given by $c(x, y)$. The problem is therefore to minimize $\int_X c\big(x, T(x)\big)\, \mathrm{d}\mu(x)$ on maps $T : X \to Y$ such that $T_*\mu = \nu$. A difficulty is that this problem might be ill posed. For example, if μ is a Dirac mass and ν is not, there is no map T such that $T_*\mu = \nu$. Moreover, even when such maps T exist, the set of such maps does not verify good properties that allow to apply classical variational methods.

Kantorovitch's tour de force [145] is twofold. He starts by relaxing the Monge problem in looking for transport plans. Those are probability measures γ on $X \times Y$ whose marginals are given by $\pi_{1*}\gamma = \mu$ and $\pi_{2*}\gamma = \nu$. He then wants to minimize $\int_{X\times Y} c(x, y)\, \mathrm{d}\gamma(x, y)$ amongst such plans. The set of plans is always nonempty (as $\mu \otimes \nu$ is one such) and it is closed and convex when X and Y are compact for instance. Hence existence of an optimal (minimizing) plan can easily be proved under mild regularity hypotheses on c. The second aspect of his contribution is to provide a dual equivalent problem. The minimal cost of a transport plan is given by

$$\sup\left(\int_Y \varphi(y)\, \mathrm{d}\nu(y) - \int_X \psi(x)\mathrm{d}\mu(x)\right),$$

where the supremum is taken amongst pairs of continuous functions $\varphi : Y \to \mathbb{R}$ and $\psi : X \to \mathbb{R}$ such that

$$\forall (x, y) \in X \times Y, \quad \varphi(y) - \psi(x) \leqslant c(x, y).$$

It will become clear in the next section that the minimizing problem of transport plans resembles that of minimizing Mather measures in Aubry–Mather theory. The dual problem of finding optimal Monge–Kantorovitch pairs of functions is transparently similar to the notion of subsolutions in weak KAM theory, especially when both spaces X and Y coincide. Finding a transport map T amounts to proving that an optimal plan is concentrated on the graph of a function from X to Y. This is analogous to Mather's Graph Theorem stating that Mather measures are concentrated on a graph. This deep parallel was drawn and studied by Bernard and Buffoni [47–49].

1.2.5 Discrete Weak KAM Theory, with an Economical Twist

The main idea of discrete weak KAM Theory is to directly discretize the Lax–Oleinik semigroup, allowing the time to take integer values only (or integer multiples of a given fixed value). This simple idea then allows to weaken hypotheses on the phase space. More precisely, coming back to the definition of the Lax–Oleinik semigroup (1.1) for $t = 1$, setting

$$h_1(x, y) = \inf_{\substack{\gamma:[-t,0]\to M \\ \gamma(0)=x \\ \gamma(1)=y}} \int_0^1 L\big(\gamma(s), \dot\gamma(s)\big)\mathrm{d}s,$$

the operator can be rewritten $U(1, x) = \inf_{y\in X} u_0(y) + h_1(y, x)$. To write such a formula, very little structure is needed. Only an underlying space X that we will assume to be metric and a function $c : X \times X \to \mathbb{R}$ that will play the role of h_1. A first theoretical study of such an operator was done in [49]. Further analogies and results coming from classical weak KAM theory were the subject of the author's PhD thesis in which he also highlighted some fundamental differences.

Here is an economical interpretation of discrete weak KAM theory. Assume that X is the (uncountable) metric space whose points are wine stores in France. Let $DRC : X \to \mathbb{R}$ be the function that gives the price $DRC(x)$ of a bottle of Domaine de la Romanée Conti in the store x.[1] Let now $c : X \times X \to \mathbb{R}$ be the function where $c(x, y)$ is the price for a 24 hour delivery of a wine bottle from x to y. Then if

[1] Domaine de la Romanée Conti is maybe the best (as far as comparisons can be made between *œuvres d'art*) and certainly most prestigious wine in the world and it is a dream of the author to taste wines from this estate in his life. Unfortunately, in this randomly chosen example, the

Maxime lives at x, the least price he will have to pay to obtain a bottle of Romanée Conti tomorrow is

$$T^- DRC(x) = \inf_{y \in X} DRC(y) + c(y, x).$$

In this simple and simplistic model, by iterating the previous operator T^-, one obtains the best price to have a bottle if one is willing to wait a long time. Finally, studying long optimal trajectories that a bottle will follow before reaching the patient Maxime will provide important objects of weak KAM theory and Aubry–Mather theory.

1.3 Organisation of the Text

The first 3 Chapters are dedicated to presenting discrete weak KAM theory in a general setting. Each of them ends with a section where related classical weak KAM results are stated to give the reader an overview of the classical theory without proofs. We believe that this back and forth between the discrete and the classical weak KAM theories is original. For well chosen costs, it highlights in a new way strong similarities and also key differences between the two versions of weak KAM theory. Discrete weak KAM theory as presented in these Chapters was developed by the author in [201, 202] (following earlier works as [49]) in a non compact setting. Here we present the compact setting that is less technical, therefore easier for a first encounter with weak KAM theory. Yet all key features and ideas of the classical theory persist and are better highlighted. Further and more precise results written hereafter were obtained with coauthors, references being provided in the text. Those first Chapters should be accessible without any specialized background.

The First Chapter introduces the Lax–Oleinik semigroups (negative and positive). The discrete weak KAM Theorem is proved and the last part of the Chapter is dedicated to constructing continuous strict subsolutions, which are a fundamental tool in the theory. This allows to define the Aubry set. In the last section, a proof of the weak KAM Theorem using the discrete one is given. Also, it is shown that for a natural cost function, weak KAM solutions and discrete weak KAM solutions coincide. The result is new to our knowledge. We also establish that for this cost, the projected Aubry set is equal to the classical one.

The Second Chapter aims at showing results of a more dynamical nature about the Aubry set. We introduce Peierls' barrier and characterize points of the projected Aubry set with it. Examples of points and chains of the Aubry set are given. Finally the problem of regularity of subsolutions (or lack thereof) is addressed. In a first part,

function DRC is close to being $+\infty$ everywhere but at Vosne-Romanée and accessible at $x =$ Vosne-Romanée (but with no available stock).

in the general setting, we characterize the projected Aubry set as the set where all subsolutions are continuous. Then by adding structure, we show existence of $C^{1,1}$ subsolutions, thus obtaining results similar to Bernard's classical ones. Finally we provide Graph Theorems and by adding a twist condition (which replaces convexity) we show how to define a partial dynamics on the Aubry set.

The Third Chapter is dedicated to Mather measures and to the crucial role they play in proving convergence of the discounted solutions. We start by giving two definitions of Mather measures and showing that they are equivalent. Then we prove convergence of solutions to the discounted equations. It is pointed out that the limit weak KAM solution for the positive Lax–Oleinik semigroup is not necessarily the conjugate of the limit weak KAM solution for the negative Lax–Oleinik semigroup. We then study a degenerate discounted problem that is new in this setting and prove convergence of thc solutions.

The Fourth Chapter provides examples in dimension 1. Those examples come from autonomous Hamiltonians and have the great merit that explicit computations can be made. We also show that the weak KAM solutions selected by the discounted approximation procedure may differ in the discrete setting and in the classical one. Such examples are folklore to experts. However we do not know of any reference where a detailed analysis is made under the scope of weak KAM theory. We believe that they are useful to have in mind in order to develop an intuition and test conjectures.

The Fifth and last Chapter puts back in the context of discrete weak KAM theory the foundational problem of conservative twist maps of the annulus. We revisit results of Mather, Aubry, Bangert . . . from the perspective of weak KAM solutions. In this unified setting we gather proofs of well known results that are spread in various references. We hope this will make them more accessible. We also give a precise description of what those weak KAM solutions look like in this setting (results of Arnaud and the author). Finally we conclude with statements of results or Arnaud and the author opening to the world of weakly integrable twist maps.

Chapter 2
The Discrete Setting, Weak KAM Solutions and Subsolutions

Abstract In this Chapter, we introduce the metric setting suitable to develop discrete weak KAM theory. We define the discrete Lax-Oleinik operator, and its discounted counterpart and prove the discrete weak KAM Theorem. We go on to introduce the fundamental Aubry set and its links to subsolutions and strict subsolutions.

Finally, we present, without proofs, the analogous results and notions coming from classical weak KAM theory. We prove new links between the classical and the discrete theories, when the cost function is given by the action functional.

The main idea of the discrete setting we focus on is to directly discretize the Lax–Oleinik semigroup, allowing the time to take integer values only (or integer multiples of a given fixed time step). This simple idea then allows to weaken the assumptions on the phase space. Most results and proofs of this Chapter are extracted as a particular compact case of [202].

2.1 Discrete Setting and the Lax–Oleinik Semigroup

We focus our attention on the case where (X, d) is a compact metric space. The analogue of the Lagrangian function is a cost function $c : X \times X \to \mathbb{R}$ which is assumed to be continuous. One can then define the Lax–Oleinik semigroup acting on the set of bounded functions $\mathcal{B}(X, \mathbb{R})$:

Definition 2.1.1 The Lax–Oleinik semigroup $T^- : \mathcal{B}(X, \mathbb{R}) \to \mathcal{B}(X, \mathbb{R})$ associates to $f : X \to \mathbb{R}$ the function

$$T^- f : x \in X \mapsto T^- f(x) = \inf_{y \in X} f(y) + c(y, x).$$

M. Zavidovique, *Discrete Weak KAM Theory*, Lecture Notes in Mathematics 2377,
https://doi.org/10.1007/978-3-031-96809-9_2

Remark 2.1.2 The Lax–Oleinik semigroup is a convolution with kernel c in the (min, +) semiring. The inf plays the role of integration and the + plays the role of multiplication in a classical convolution.

In particular, if the set X is finite, the Lax–Oleinik semigroup reduces to a product (in the (min, +) semiring) with the matrix whose entries are given by c.

If f is a continuous function, then the infimum in the definition of $T^- f(x)$ is a minimum by compactness of X.

We define the sup-norm $\|\cdot\|_\infty$ on the space $\mathcal{B}(X, \mathbb{R})$ by setting $\|f\|_\infty = \sup_{x\in X} |f(x)|$ for $f \in \mathcal{B}(X, \mathbb{R})$. The normed vector space $\big(\mathcal{B}(X, \mathbb{R}), \|\cdot\|_\infty\big)$ is a Banach space.

We start by listing first basic properties of T^-:

Proposition 2.1.3

(i) The image of T^- consists of equicontinuous functions with uniformly bounded amplitude.[1]
(ii) The Lax–Oleinik semigroup commutes with addition of constant functions: if $k \in \mathbb{R}$ and f is a function, $T^-(f + k) = (T^- f) + k$.
(iii) The Lax–Oleinik semigroup is order preserving: if $f \leqslant g$ then $T^- f \leqslant T^- g$.
(iv) The Lax–Oleinik semigroup is 1-Lipschitz for the sup-norm $\|\cdot\|_\infty$.

Proof Let us consider a modulus of uniform continuity ω for c (X being compact). This is a non-decreasing function $\omega : [0, +\infty) \to [0, +\infty)$ that is continuous at 0, with $\omega(0) = 0$, such that

$$\forall (x, y, x', y') \in X^4, \quad |c(x, y) - c(x', y')| \leqslant \omega\big(d(x, x') + d(y, y')\big).$$

Without loss of generality, by triangular inequality we may assume that ω is bounded and that $\|\omega\|_\infty \leqslant 2\|c\|_\infty$. Let $f : X \to \mathbb{R}$ be a bounded function and $\varepsilon > 0$. Let $(x, x') \in X^2$. By definition of the Lax–Oleinik semigroup, there exists a y_ε such that $T^- f(x') \geqslant f(y_\varepsilon) + c(y_\varepsilon, x') - \varepsilon$. It follows that

$$T^- f(x) - T^- f(x') \leqslant f(y_\varepsilon) + c(y_\varepsilon, x) - f(y_\varepsilon) - c(y_\varepsilon, x') + \varepsilon \leqslant \omega\big(d(x, x')\big) + \varepsilon.$$

Letting $\varepsilon \to 0$ yields that $T^- f(x) - T^- f(x') \leqslant \omega\big(d(x, x')\big)$. As x and x' play symmetrical roles we find that $|T^- f(x) - T^- f(x')| \leqslant \omega\big(d(x, x')\big)$. This is (i) as x, x' are arbitrary.

Points (ii) and (iii) are obvious from the definition of T^- and automatically imply (iv). Indeed, if f and g are bounded functions, as

$$f - \|f - g\|_\infty \leqslant g \leqslant f + \|f - g\|_\infty,$$

[1] By amplitude of a function $f : X \to \mathbb{R}$ we mean $\sup f - \inf f$.

we obtain

$$T^- f - \|f - g\|_\infty \leqslant T^- g \leqslant T^- f + \|f - g\|_\infty$$

which means that $\|T^- f - T^- g\|_\infty \leqslant \|f - g\|_\infty$. □

Remark 2.1.4 The Lax–Oleinik semigroup can actually be defined on arbitrary functions $f : X \to \mathbb{R}$ (not necessarily bounded) with the only modification that $T^- f$ can take the value $-\infty$. However, it can be easily checked that, as c is bounded, this may only happen if f is unbounded from below. In this case, $T^- f$ is identically $-\infty$. Otherwise, the conclusions of the previous Proposition 2.1.3 still hold, with the same proofs.

Now that those properties have been established, let us move on to the weak KAM theorem.

2.2 The Weak KAM Theorem and Critical Subsolutions

In this section, we will introduce and construct some of the most important objects of weak KAM theory. The first ones are of course weak KAM solutions and are given by the following theorem:

Theorem 2.2.1 (Weak KAM) *There exists a unique constant* $c[0] \in \mathbb{R}$ *for which the equation* $u = T^- u + c[0]$ *admits solutions* $u : X \to \mathbb{R}$.

Remark 2.2.2 Such functions are then called *weak KAM solutions*. The constant $c[0]$ is called the *critical value*.

It is immediate from Proposition 2.1.3 and Remark 2.1.4 that weak KAM solutions are automatically continuous.

We will give two proofs of the existence part of the weak KAM Theorem. The first one is similar to the original proof of Fathi [105, 109]. The second one is reminiscent of the work of Lions et al. [152] on homogenization, that actually appeared prior to the work of Fathi.

The uniqueness of the constant $c[0]$ will be established in a second step.

First Proof Let us introduce $\mathcal{E} = \mathcal{B}(X, \mathbb{R})/\mathbb{R}\mathbb{1}$ the quotient vector space of bounded functions by constant functions. The set $\mathcal{E}$ is clearly a vector space. If f is a bounded function, we will denote by $\bar{f}$ its projection in $\mathcal{E}$. There is an induced norm on $\mathcal{E}$: if $f : X \to \mathbb{R}$ is a continuous function, denoting $\bar{f} \in \mathcal{E}$ its class in the quotient, we set $\|\bar{f}\|_0 = \min\{\|f + k\|_\infty,\quad k \in \mathbb{R}\}$. As T^- commutes with addition of constant functions, it induces an operator $\mathcal{T}$ on $\mathcal{E}$ defined by $\mathcal{T}\bar{f} = \overline{T^- f}$ which is independent of the representative f in the equivalence class $\bar{f}$. This new operator is also continuous. Indeed, if f, g are two bounded functions, then for some suitably chosen constant $k \in \mathbb{R}$,

$$\|\mathcal{T}\bar{f} - \mathcal{T}\bar{g}\|_0 \leqslant \|T^- f - T^- g + k\|_\infty \leqslant \|f - g + k\|_\infty = \|\bar{f} - \bar{g}\|_0.$$

Moreover, it follows from the fact that T^- has values in equicontinuous functions with uniformly bounded amplitude (see Proposition 2.1.3) and from the Arzelà–Ascoli theorem [99, Theorem 6.4 page 267] that $\mathcal{T}(\mathcal{E})$ is relatively compact. We can therefore apply the Schauder–Tychonoff Theorem [99, Theorems 2.2 and 3.2 pages 414-415] which asserts that $\mathcal{T}$ has a fixed point. This exactly means that there exists a bounded function $u : X \to \mathbb{R}$ and a constant C such that $u = T^- u + C$. □

The use of the Schauder–Tychonoff Theorem at the end of this first proof, though natural, is not really necessary. Indeed, fixed points of 1-Lipschitz maps can be obtained by much simpler arguments, usually by perturbing the map, making it contracting and then passing to the limit. This is the spirit of the second proof in which we use an approximation called *discounted procedure*:

Second Proof Let $\lambda \in (0, 1)$ and let us introduce the discounted operator T_λ^- which acts on bounded functions as follows:

$$\forall f \in \mathcal{B}(X, \mathbb{R}), \quad \forall x \in X, \quad T_\lambda^- f(x) = \inf_{y \in X} \lambda f(y) + c(y, x) = T^-(\lambda f)(x).$$

Of course, the last formulation, together with the 1-Lipschitz nature of T^- imply that T_λ^- is now λ-Lipschitz. Hence, by the Banach fixed point theorem [99, Theorem 7.2 page 305], as $\mathcal{B}(X, \mathbb{R})$ is a Banach space, T_λ^- admits a unique fixed point u_λ which then verifies $T_\lambda^- u_\lambda = u_\lambda$. By Proposition 2.1.3, the $(u_\lambda)_{\lambda \in (0,1)}$ are equicontinuous with uniformly bounded amplitude, as they all belong to the image of T^-. Moreover

$$\forall x \in X, \qquad \min_{(y,z) \in X \times X} c(y, z) \leqslant (1 - \lambda) u_\lambda(x) \leqslant \max_{z \in X} c(z, z).$$

To prove the left inequality, fix λ and take x_1 such that $u_\lambda(x_1)$ is minimal. One then has for some $y \in X$,

$$u_\lambda(x_1) = \lambda u_\lambda(y) + c(y, x_1) \geqslant \lambda u_\lambda(x_1) + c(y, x_1).$$

The right inequality follows from the fact that by definition of T_λ^- we obviously have $u_\lambda(x) = T_\lambda^- u_\lambda(x) \leqslant \lambda u_\lambda(x) + c(x, x)$.

Let us fix a point $x_0 \in X$ and define $\hat{u}_\lambda = u_\lambda - u_\lambda(x_0)$ for all $\lambda \in (0, 1)$. The previous remarks show that we can find a sequence $\lambda_n \to 1$ such that $(1 - \lambda_n) u_{\lambda_n}(x_0)$ converges to a constant C and $(\hat{u}_{\lambda_n})_{n \in \mathbb{N}}$ uniformly converges to a function u, as $n \to +\infty$. Note that in fact, as the functions u_λ have bounded amplitude, one has $(1-\lambda_n)u_{\lambda_n} \to C$ uniformly. It now follows that

$$\begin{aligned} \forall x \in X, \quad \hat{u}_{\lambda_n}(x) &= u_{\lambda_n}(x) - u_{\lambda_n}(x_0) \\ &= T^-(\lambda_n u_{\lambda_n})(x) - u_{\lambda_n}(x_0) \\ &= T^-(\lambda_n \hat{u}_{\lambda_n})(x) + (\lambda_n - 1)u_{\lambda_n}(x_0). \end{aligned}$$

Letting $n \to +\infty$, by continuity of T^-, we conclude that $u = T^-u - C$. □

It remains to prove the uniqueness of the critical constant $c[0]$. This is a direct consequence of the following proposition:

Proposition 2.2.3 *Assume that* $u = T^-u + c[0]$ *for some function* u *and constant* $c[0]$, *then*

$$\forall v \in \mathcal{B}(X,\mathbb{R}), \qquad \frac{T^{-n}v}{n} \underset{n\to+\infty}{\longrightarrow} -c[0],$$

where $T^{-n} = T^- \circ \cdots \circ T^-$ *denotes the* n*-th iterate of the Lax–Oleinik semigroup, and the convergence is uniform.*

Proof This result is a direct consequence of the non expansive character of the Lax–Oleinik semigroup. Indeed, by induction, one obtains that for all integers $n > 0$, $u = T^{-n}u + nc[0]$. Hence, if $v \in \mathcal{B}(X,\mathbb{R})$, then

$$\|T^{-n}u - T^{-n}v\|_\infty = \|u - nc[0] - T^{-n}v\|_\infty \leqslant \|u - v\|_\infty.$$

The result follows immediately as $\|nc[0] + T^{-n}v\|_\infty \leqslant \|u - v\|_\infty + \|u\|_\infty$. □

Remark 2.2.4

(i) In the previous Proposition, one derives that for all $v \in \mathcal{B}(X,\mathbb{R})$ the sequence $(T^{-n}v + nc[0])_{n\in\mathbb{N}}$ is uniformly bounded.
(ii) It follows from the uniqueness of $c[0]$ that, in the second proof of the weak KAM Theorem (2.2.1), the whole family $(1-\lambda)u_\lambda$ uniformly converges to $c[0]$ as $\lambda \to 1$.
(iii) If $u : X \to \mathbb{R}$ is a weak KAM solution, it satisfies the following fundamental inequalities:

$$\forall (x,y) \in X \times X, \quad u(y) - u(x) \leqslant c(x,y) + c[0].$$

This last Remark motivates the following definition:

Definition 2.2.5 Given $C \in \mathbb{R}$, a function $v : X \to \mathbb{R}$ will be termed *C-subsolution* if

$$\forall (x, y) \in X \times X, \quad v(y) - v(x) \leqslant c(x, y) + C,$$

or equivalently, if $v \leqslant T^- v + C$.

We will call $c[0]$-subsolutions critical subsolutions, or just subsolutions when no confusion is possible. We will denote by $\mathcal{S}_C$ the set of C-subsolutions, and by $\mathcal{S} = \mathcal{S}_{c[0]}$ the set of critical subsolutions.

Here are some first properties of subsolutions:

Proposition 2.2.6 *Given $C \in \mathbb{R}$, the following hold:*

(i) Any C-subsolution is bounded.
(ii) The set $\mathcal{S}_C$ of C-subsolutions is closed (with respect to pointwise convergence), convex and stable by the Lax–Oleinik semigroup: $T^-(\mathcal{S}_C) \subset \mathcal{S}_C$.

Proof The first point comes from the fact that if $u \in \mathcal{S}_C$ and $x_0 \in X$, for all $y \in X$,

$$u(x_0) - c(y, x_0) - C \leqslant u(y) \leqslant u(x_0) + c(x_0, y) + C,$$

hence $\|u\|_\infty \leqslant |u(x_0)| + \|c + C\|_\infty$.

The fact that $\mathcal{S}_C$ is closed and convex is immediate from the definition. Only stability deserves an explanation. It follows from the fact that $u \in \mathcal{S}_C$ if and only if $u \leqslant T^- u + C$ as can be checked by applying the definitions. Consequently, by the properties of the Lax–Oleinik semigroup (Proposition 2.1.3), if $u \leqslant T^- u + C$ then

$$T^- u \leqslant T^-(T^- u + C) = T^-(T^- u) + C,$$

which means $T^- u \in \mathcal{S}_C$. □

We conclude this section by one last characterization of the critical constant $c[0]$:

Proposition 2.2.7 *The following holds:* $c[0] = \min\{C \in \mathbb{R}, \ \ \mathcal{S}_C \neq \varnothing\}$.

Proof As follows from the weak KAM Theorem, $\mathcal{S}_{c[0]} \neq \varnothing$, so we just have to prove that if for some constant C, $\mathcal{S}_C \neq \varnothing$, then $C \geqslant c[0]$. Let then $u \in \mathcal{S}_C$ for some $C \in \mathbb{R}$. As in the last proof, we get by induction that for all positive integer n, $u \leqslant T^{-n} u + nC$. Therefore, dividing by n, we infer that $u/n - C \leqslant T^{-n} u/n$. But the right hand side converges to $-c[0]$ by Proposition 2.2.3. Hence passing to the limit we conclude that $-C \leqslant -c[0]$. □

We illustrate once more our model with a random liquid example as in the introductory section page 9. Here again, X is the space of wine stores in France and $c : X \times X \to \mathbb{R}$ provides the cost $c(x, y)$ of bringing a bottle of wine from x to

y in a day. Let $P : X \to \mathbb{R}$ denote the price $P(x)$ of a bottle of Petrus[2] at a location x. Let $n \in \mathbb{N}$, then if the author is at $x \in X$, the best price he will pay to have a bottle of Petrus in n days is $T^{-n}P(x)$. Proposition 2.2.3 then states that whatever the initial price P, for $n \to +\infty$ the amount $T^{-n}P(x)$ grows like $-nc[0]$. In this particular example, given the order of magnitude of P, the time n would have to be very very large to compensate the initial price.

2.3 The Positive Lax–Oleinik Semigroup

In this section, we introduce the positive Lax–Oleinik semigroup and state its main properties. As they are analogous to the properties of the (negative) Lax–Oleinik semigroup, the proofs are omitted and left to the reader.

Definition 2.3.1 The positive Lax–Oleinik semigroup $T^+ : \mathcal{B}(X, \mathbb{R}) \to \mathcal{B}(X, \mathbb{R})$ maps to $f : X \to \mathbb{R}$ the function

$$T^+f : x \in X \mapsto T^+f(x) = \sup_{y \in X} f(y) - c(x, y).$$

Remark 2.3.2 From the definition, one checks that $T^+f = -\big(T^-_{\hat{c}}(-f)\big)$ where $T^-_{\hat{c}}$ is the cost defined by $\hat{c}(x, y) = c(y, x)$. Hence the fact that T^- and T^+ share very similar properties is not surprising.

Proposition 2.3.3

(i) The image of T^+ consists of equicontinuous functions with uniformly bounded amplitude.
(ii) The positive Lax–Oleinik semigroup, T^+, commutes with addition of constant functions.
(iii) The positive Lax–Oleinik semigroup is order preserving:
if $f \leqslant g$ then $T^+f \leqslant T^+g$.
(iv) The positive Lax–Oleinik semigroup is 1-Lipschitz for the sup-norm $\|\cdot\|_\infty$.

The positive semigroup also fulfills a weak KAM Theorem:

Theorem 2.3.4 (Positive Weak KAM) *The critical constant $c[0]$ is the unique real value $c \in \mathbb{R}$ for which the equation $u = T^+u - c$ admits solutions $u : X \to \mathbb{R}$. Moreover the constant $c[0]$ has the following characterization:*

$$\forall v \in \mathcal{B}(X, \mathbb{R}), \qquad \frac{T^{+n}v}{n} \underset{n \to +\infty}{\longrightarrow} c[0],$$

[2] Petrus is a renowned and quite inaccessible wine from the Bordeaux region in France. More precisely it is the leading estate of the appellation Pomerol. The royalties (or lack thereof) earned by a lifetimes'work of the author would probably allow him to buy a quarter of a bottle of Petrus.

where $T^{+n} = T^+ \circ \cdots \circ T^+$ *denotes the n-th iterate of the positive Lax–Oleinik semigroup, and the convergence is uniform.*

Remark 2.3.5 Such functions are then called *positive weak KAM solutions.*

It is immediate from Proposition 2.3.3 that positive weak KAM solutions are always continuous.

In the positive weak KAM theorem, the fact that the critical constant for T^+ is the same as that of T^-, that is $c[0]$, deserves some explanation. It is checked that $u : X \to \mathbb{R}$ is a C-subsolution, for some $C \in \mathbb{R}$ if and only if $u \geqslant T^+u - C$. Hence, as for the Lax–Oleinik semigroup, one has that $T^+(\mathcal{S}_C) \subset \mathcal{S}_C$.

Now, the fact that both critical constants coincide follows from Proposition 2.2.7 which characterizes the critical constant only making use of the notion of subsolution, and not of either the positive, nor negative Lax–Oleinik semigroup. However, its proof can be done equivalently using the negative or the positive Lax–Oleinik semigroup.

We end this section with a curiosity on the composition of positive and negative Lax–Oleinik semigroups:

Proposition 2.3.6

1. *Let* $u \in \mathcal{B}(X, \mathbb{R})$, *then* $T^+ \circ T^-u \leqslant u$ *and* $T^- \circ T^+u \geqslant u$.
2. *Let* $u \in \mathcal{B}(X, \mathbb{R})$, *then* $T^- \circ T^+ \circ T^-u = T^-u$ *and* $T^+ \circ T^- \circ T^+u = T^+u$.

Proof

1. Let us establish the first inequality. If $x \in X$ then

$$T^+ \circ T^-u(x) = \sup_{y \in X} T^-u(y) - c(x, y) = \sup_{y \in X} \inf_{z \in X} u(z) + c(z, y) - c(x, y) \leqslant u(x),$$

where the last inequality is obtained by taking $z = x$.

2. We again establish only the first equality. We have, using both inequalities of the first part, $(T^- \circ T^+)(T^-u) \geqslant T^-u$ and $T^-(T^+ \circ T^-u) \leqslant T^-u$.

□

2.4 Strict Subsolutions, Aubry Sets

In this section, we will focus our study on critical subsolutions, hence the adjective may be omitted from time to time, but is always implicit. Recall that the set $\mathcal{S}$ is the set of critical subsolutions. The goal will be to construct a special kind of subsolutions which are in some sense better than the others:

Theorem 2.4.1 *There exists a subsolution* $u_0 \in \mathcal{S} \cap C^0(X, \mathbb{R})$ *such that, if the equality* $u_0(y) - u_0(x) = c(x, y) + c[0]$ *holds for some* $(x, y) \in X \times X$, *then*

$$\forall u \in \mathcal{S}, \quad u(y) - u(x) = c(x, y) + c[0]. \tag{2.1}$$

A subsolution verifying this last property (2.1) will be termed *strict*. The proof of this Theorem will occupy the rest of this section. It will be divided in two main parts. In a first one, we will construct the function u_0 and prove that it verifies (2.1) for all other continuous subsolutions.

Then a digression is devoted to studying the structure of the set where the equality $u_0(y) - u_0(x) = c(x, y) + c[0]$ takes place. This is the Aubry set, a central object in weak KAM theory, and unsurprisingly in Aubry–Mather theory.

This being achieved, the proof of the Theorem ends rather easily.

Beginning of the Proof The set (X, d) being metric and compact, the Banach space $\big(C^0(X, \mathbb{R}), \|\cdot\|_\infty\big)$ is itself separable. Therefore, $\mathcal{S} \cap C^0(X, \mathbb{R})$ being a subset of a separable space is also separable. Let $(v_n)_{n\in\mathbb{N}}$ be a dense (with respect to the topology of uniform convergence) sequence in $\mathcal{S} \cap C^0(X, \mathbb{R})$. Now, let us define $u_0 = \sum\limits_{n\geqslant 0} a_n v_n$ where for $n > 0$, we set $a_n = \min(2^{-n}, 2^{-n}/\|v_n\|_\infty)$ and $a_0 = 1 - \sum\limits_{n>0} a_n$.

The function u_0 is defined by a series of continuous functions that converges for the $\|\cdot\|_\infty$-norm, hence it is continuous. It is an infinite convex combination of critical subsolutions, therefore, by Proposition 2.1.3 it is a subsolution. We will prove it verifies the property we seek for.

Let us consider now $(x, y) \in X \times X$ such that $u_0(y) - u_0(x) = c(x, y) + c[0]$. By definition of subsolutions, we know that $v_n(y) - v_n(x) \leqslant c(x, y) + c[0]$, for all $n \geqslant 0$. Multiplying each of these inequalities by a_n and summing, we get

$$u_0(y) - u_0(x) = \sum_{n=0}^{+\infty} a_n\big(v_n(y) - v_n(x)\big) \leqslant \sum_{n=0}^{+\infty} a_n(c(x, y) + c[0]) = c(x, y) + c[0],$$

which is in fact an equality. Hence all the middle inequalities must be equalities, and the a_n being all positive, we conclude that $v_n(y) - v_n(x) = c(x, y) + c[0]$ for all integer n. The sequence $(v_n)_{n\in\mathbb{N}}$ being dense in $\mathcal{S} \cap C^0(X, \mathbb{R})$ we conclude eventually that

$$\forall u \in \mathcal{S} \cap C^0(X, \mathbb{R}), \quad \Big[u_0(y) - u_0(x) = c(x, y) + c[0]\Big] \\ \Longrightarrow \Big[u(y) - u(x) = c(x, y) + c[0]\Big].$$

□

The rest of the proof will consist in extending this property to non continuous subsolutions.

Let us start by defining some useful sets:

Definition 2.4.2 (Aubry Sets) Let $u \in \mathcal{S}$ be a critical subsolution.

- The non-strict set of u is

$$\mathcal{NS}_u = \{(x, y) \in X \times X, \quad u(y) - u(x) = c(x, y) + c[0]\}.$$

- The Aubry set of u is

$$\widetilde{\mathcal{A}}_u = \Big\{(x_n)_{n\in\mathbb{Z}} \in X^{\mathbb{Z}}, \ \forall n < p, \quad u(x_p) - u(x_n) = \sum_{k=n}^{p-1} c(x_k, x_{k+1}) + (n-p)c[0]\Big\}.$$

- The Aubry set is $\widetilde{\mathcal{A}} = \widetilde{\mathcal{A}}_{u_0} \subset X^{\mathbb{Z}}$ where u_0 is the peculiar subsolution previously constructed.
- The 2-Aubry is $\widehat{\mathcal{A}} = \mathcal{NS}_{u_0} \subset X \times X$.
- Finally, the projected Aubry set is $\mathcal{A} = \pi_1(\widehat{\mathcal{A}}) \subset X$ where $\pi_1 : X \times X \to X$ is the projection on the first factor: $(x, y) \mapsto x$.

Remark 2.4.3

(i) The beginning of the proof of Theorem 2.4.1 may be summed up in the equality $\widehat{\mathcal{A}} = \bigcap\limits_{u\in\mathcal{S}\cap C^0(X,\mathbb{R})} \mathcal{NS}_u$.

(ii) If $s : X^{\mathbb{Z}} \to X^{\mathbb{Z}}$ is the shift operator: $(x_n)_{n\in\mathbb{Z}} \mapsto (x_{n+1})_{n\in\mathbb{Z}}$ then the sets introduced above are invariant by this shift: $\widetilde{\mathcal{A}}_u = s(\widetilde{\mathcal{A}}_u)$ and $\widetilde{\mathcal{A}} = s(\widetilde{\mathcal{A}})$.

(iii) If $u \in \mathcal{S}$ and $(x_n)_{n\in\mathbb{Z}} \in X^{\mathbb{Z}}$ then one has for all integers $k \in \mathbb{Z}$,

$$u(x_{k+1}) - u(x_k) \leqslant c(x_k, x_{k+1}) + c[0].$$

Summing these inequalities, one gets

$$\forall n < p, \quad u(x_p) - u(x_n) \leqslant \sum_{k=n}^{p-1} c(x_k, x_{k+1}) + (n-p)c[0].$$

Therefore if for some $n < p$, the previous inequality turns out to be an equality: $u(x_p) - u(x_n) = \sum\limits_{k=n}^{p-1} c(x_k, x_{k+1}) + (n-p)c[0]$, it implies the equalities

$$\forall n \leqslant k < p, \quad u(x_{k+1}) - u(x_k) = c(x_k, x_{k+1}) + c[0].$$

More generally, under the same hypothesis, if $n \leqslant n' < p' \leqslant p$, then

$$u(x_{p'}) - u(x_{n'}) = \sum_{k=n'}^{p'-1} c(x_k, x_{k+1}) + (n' - p')c[0].$$

(iv) Note that by taking $n = 0$ and $p = 1$ in the definition of $\widetilde{\mathcal{A}}_u$ we find that $\pi_{0,1}(\widetilde{\mathcal{A}}_u) \subset \mathcal{NS}_u$ where $\pi_{0,1} : X^{\mathbb{Z}} \to X \times X$ is the projection: $(x_n)_{n\in\mathbb{Z}} \mapsto (x_0, x_1)$.

Now let us study these Aubry sets. We start by basic topological properties:

Proposition 2.4.4 *The 2-Aubry set $\widehat{\mathcal{A}}$ is closed and non-empty.*

Proof Being closed comes from the identity $\widehat{\mathcal{A}} = F^{-1}\{0\}$ where $F(x, y) = u_0(y) - u_0(x) - c(x, y) - c[0]$ is continuous.

Being non-empty is a consequence of the minimality of $c[0]$ (Proposition 2.2.7). Indeed, as u_0 is a subsolution, the function F is non-positive. By compactness and continuity, if $\widehat{\mathcal{A}} = F^{-1}\{0\}$ were empty, there would be a small $\varepsilon > 0$ such that $u_0(y) - u_0(x) \leqslant c(x, y) + c[0] - \varepsilon$, for all $(x, y) \in X \times X$. But this means that $u_0 \in \mathcal{S}_{c[0]-\varepsilon} \neq \varnothing$ which contradicts Proposition 2.2.7. □

The next proposition states that elements of the 2-Aubry set come in families, meaning that the Aubry set is not empty:

Proposition 2.4.5 *Let $(x, y) \in \widehat{\mathcal{A}}$, then there exists a sequence $(x_n)_{n\in\mathbb{Z}} \in \widetilde{\mathcal{A}}$ such that $(x, y) = \pi_{0,1}\big((x_n)_{n\in\mathbb{Z}}\big)$. In particular, the Aubry set is itself closed (for the product topology), not empty, and $\widehat{\mathcal{A}} = \pi_{0,1}(\widetilde{\mathcal{A}})$.*

The proof will make use of the following two lemmas. The first one's proof is a direct application of the definitions and is omitted:

Lemma 2.4.6 *Let n be a positive integer and $f \in \mathcal{B}(X, \mathbb{R})$, then*

$$\forall x \in X, \quad T^{-n} f(x) = \inf_{x_{-n},\cdots,x_0=x} f(x_{-n}) + \sum_{k=-n}^{-1} c(x_k, x_{k+1}),$$

$$\forall x \in X, \quad T^{+n} f(x) = \sup_{x=x_0,\cdots,x_n} f(x_n) - \sum_{k=0}^{n-1} c(x_k, x_{k+1}).$$

Lemma 2.4.7 *Let $u \in \mathcal{S}$ be a continuous subsolution and $(x, y) \in \widehat{\mathcal{A}}$. We have $T^{-n}u(x) = u(x) - nc[0]$ and $T^{+n}u(y) = u(y) + nc[0]$ for all positive integer n.*

Proof As $T^{-n}u$ and $T^{+n}u$ are continuous subsolutions and $(x, y) \in \widehat{\mathcal{A}}$, we now know that $T^{-n}u(y) - T^{-n}u(x) = T^{+n}u(y) - T^{+n}u(x) = c(x, y) + c[0]$. It implies readily that

$$T^{-(n+1)}u(y) + c[0] \geqslant T^{-n}u(y) = T^{-n}u(x) + c(x, y) + c[0] \geqslant T^{-(n+1)}u(y) + c[0],$$

the first inequality coming from the fact that $T^{-n}u \in \mathcal{S}$ the second from the definition of T^-. Hence all inequalities turn out to be equalities and the sequence $(T^{-n}u(y) + nc[0])_{n\in\mathbb{N}}$ is constant.

The same holds for $T^{-n}u(x) + nc[0] = T^{-n}u(y) + nc[0] - c(x, y) - c[0]$.

The proof of the rest of the lemma is established similarly.

□

Proof of Proposition 2.4.5 The sequences (x_n) for $n \leqslant 0$ and $n \geqslant 0$ are constructed separately. The first half will come from the negative Lax–Oleinik semigroup, the second one, from the positive Lax–Oleinik semigroup.

As u_0 and c are continuous and X is compact, any supremum (resp. infimum) involving them is actually a maximum (resp. minimum). Therefore, for each positive n, there exist chains $x^n_{-n}, \cdots, x^n_0 = x, x^n_1 = y, \cdots x^n_{n+1}$ such that

$$T^{-n}u_0(x) = u_0(x^n_{-n}) + \sum_{k=-n}^{-1} c(x^n_k, x^n_{k+1}) \ ;$$

$$T^{+n}u_0(y) = u_0(x^n_{n+1}) - \sum_{k=1}^{n} c(x^n_k, x^n_{k+1}).$$

By a diagonal argument, let $(N_n)_{n\geqslant 0}$ be an extracted sequence such that for all $k \geqslant 0$ the sequences $(x^{N_n}_k)_{n\geqslant k}$ and $(x^{N_n}_{-k})_{n\geqslant k}$ converge. We will denote by x_k and x_{-k} the respective limits. Obviously, $x_0 = x$ and $x_1 = y$. By definition of the Lax–Oleinik semigroup and Lemma 2.4.7, for all $k \leqslant n$,

$$u_0(x) = u_0(x^{N_n}_k) + \sum_{i=-k}^{-1} c(x^{N_n}_i, x^{N_n}_{i+1}) + kc[0] \ ;$$

$$u_0(y) = u_0(x^{N_n}_{k+1}) - \sum_{i=1}^{k} c(x^{N_n}_i, x^{N_n}_{i+1}) - kc[0].$$

Letting $n \to +\infty$, we obtain that

$$u_0(x) = u_0(x_k) + \sum_{i=-k}^{-1} c(x_i, x_{i+1}) + kc[0] \ ;$$

$$u_0(y) = u_0(x_{k+1}) - \sum_{i=1}^{k} c(x_i, x_{i+1}) - kc[0].$$

As in Remark 2.4.3, this implies that for all $n \in \mathbb{Z}$, $u_0(x_{n+1}) - u_0(x_n) = c(x_n, x_{n+1}) + c[0]$ (the case $n < 0$ is given by the left equalities and $n > 0$ by the right equalities above, $n = 0$ is because $(x, y) \in \widehat{\mathcal{A}}$). Now, again as in Remark 2.4.3, by summing those equalities, one obtains that

$$\forall n < p, \quad u_0(x_p) - u_0(x_n) = \sum_{k=n}^{p-1} c(x_k, x_{k+1}) + (n - p)c[0].$$

This exactly means that $(x_n)_{n\in\mathbb{Z}} \in \widetilde{\mathcal{A}}$. □

Remark 2.4.8 Let $(x_n)_{n\in\mathbb{Z}} \in \widetilde{\mathcal{A}}$ and u be a subsolution. As for all $n \in \mathbb{Z}$, $(x_n, x_{n+1}) \in \widehat{\mathcal{A}}$, summing equalities $u(x_{n+1}) - u(x_n) = c(x_n, x_{n+1}) + c[0]$, one finds that

$$\forall n < p, \quad u(x_p) - u(x_n) = \sum_{k=n}^{p-1} c(x_k, x_{k+1}) + (n - p)c[0].$$

In other terms, $\widetilde{\mathcal{A}} \subset \widetilde{\mathcal{A}}_u$.

As the Aubry set is invariant by the shift, we get the immediate:

Corollary 2.4.9 *The projected Aubry set can be obtained by either projection, $\mathcal{A} = \pi_1(\widehat{\mathcal{A}}) = \pi_2(\widehat{\mathcal{A}})$, where π_2 is the projection on the second factor.*

Let us now complete the proof of the Theorem:

End of the Proof of Theorem 2.4.1 Let $u \in \mathcal{S}$ be any subsolution. Let $(x, y) \in \widehat{\mathcal{A}}$ and let $(x_n)_{n\in\mathbb{Z}} \in \widetilde{\mathcal{A}}$ such that $\pi_{0,1}\big((x_n)_{n\in\mathbb{Z}}\big) = (x, y)$. We have the following chain of inequalities, for all integer $n \in \mathbb{Z}$:

$$T^-u(x_{n+1}) \leqslant u(x_n)+c(x_n, x_{n+1}) \leqslant T^-u(x_n)+c(x_n, x_{n+1})+c[0] = T^-u(x_{n+1}),$$

$$T^+u(x_{n-1}) \geqslant u(x_n)-c(x_{n-1}, x_n) \geqslant T^+u(x_n)-c(x_{n-1}, x_n)-c[0] = T^+u(x_{n-1}).$$

In each line, the first inequality is by definition of the Lax–Oleinik semigroups, the second holds because $u \in \mathcal{S}$ and the last equality comes from the fact that, both T^-u and T^+u being continuous subsolutions, the first part of the proof of Theorem 2.4.1 applies.

Hence all inequalities are equalities, and taking $n = 0, 1$ it follows both equalities $u(x) = T^-u(x) + c[0]$ and $u(y) = T^-u(y) + c[0]$ and eventually that

$$u(y) - u(x) = T^-u(y) - T^-u(x) = c(x, y) + c[0].$$

This ends the proof. □

Let us conclude by a corollary of this proof:

Corollary 2.4.10 *Let $x \in X$. The following assertions are equivalent:*

(i) $x \in \mathcal{A}$,
(ii) $\forall u \in \mathcal{S},\ \ u(x) = T^- u(x) + c[0]$,
(iii) $\forall u \in \mathcal{S},\ \ u(x) = T^+ u(x) - c[0]$,
(iv) for any $u \in \mathcal{S}$ and $n > 0$, $u(x) = T^{-n} u(x) + nc[0] = T^{+n} u(x) - nc[0]$.

2.5 Relations to the Classical Theory

2.5.1 *Classical Setting and Lax–Oleinik Semigroup*

Classical weak KAM theory takes place originally in a smooth, connected and compact manifold M. We will denote by TM the tangent bundle of M and denote points in this set by $(x, v) \in TM$, where $x \in M$ and $v \in T_x M$ is a vector tangent to M at x. Similarly, T^*M is the cotangent bundle of M, and a point of this cotangent bundle will be written $(x, p) \in T^*M$, where $x \in M$ and $p \in T_x^*M$ is a linear form on $T_x M$. For convenience, we will equip TM with a Riemannian metric and denote by $(x, v) \mapsto \|v\|_x$ the associated norm. As M is compact, all Riemannian metrics are equivalent and all results are independent of this choice. The induced distance on M will be denoted by $\mathrm{d}(\cdot, \cdot)$.[3] One considers a Tonelli Hamiltonian, that is a function $H : T^*M \to \mathbb{R}$ defined on the cotangent bundle of M verifying the following set of conditions:

- H is C^2,
- H is strictly convex in the momentum variable, meaning that for any $(x, p) \in T^*M$ the Hessian $\partial_{pp} H(x, p)$ is positive definite.
- H is superlinear, meaning that

$$\forall x \in M, \qquad \lim_{\|p\|_x \to +\infty} \frac{H(x, p)}{\|p\|_x} = +\infty.$$

Note that in the superlinearity condition the limit is automatically uniform in x, thanks to the convexity of H and to the compactness of M. Moreover, this condition depends at first sight on the choice of the Riemannian metric on TM, the norm of $p \in T_x^*M$ being the operator norm of p and denoted again $\|p\|_x$ to simplify notations. However, M being compact, any two Riemannian metrics are equivalent, hence the notion of superlinearity becomes independent of the initial choice.

[3] A simple example to keep in mind is that of the flat torus $M = \mathbb{T}^n = \mathbb{R}^n/\mathbb{Z}^n$. In this case, both TM and T^*M are isomorphic to $\mathbb{T}^n \times \mathbb{R}^n$. As a Riemannian metric, one may use the canonical Euclidian scalar product on $\mathbb{R}^n$ to define a metric both on $T\mathbb{T}^n$ and on $T^*\mathbb{T}^n$.

Given this Hamiltonian, one studies two equations called *Hamilton–Jacobi equations*. The evolutionary Hamilton–Jacobi equation is:

$$\begin{cases} \partial_t u + H(x, \partial_x u) = 0, \\ u(0, \cdot) = u_0. \end{cases} \tag{EHJ}$$

Above, the unknown function is $u(t, x) : [0, +\infty) \times M \to \mathbb{R}$ and $u_0 : M \to \mathbb{R}$ is a given continuous function called initial condition.

The stationary Hamilton–Jacobi equation is

$$H(x, D_x u) = \alpha, \tag{SHJ}$$

where the unknown is $u : M \to \mathbb{R}$ and $\alpha \in \mathbb{R}$ is a given constant.

Strong (C^1) solutions to those equations rarely exist. For instance, if u_0 is smooth, one can solve the evolutionary equation by using the method of characteristics. However, shocks appear almost systematically in finite time and the solution ceases to be smooth (results about this can be found in [92, 109]). Therefore a notion of weak solutions is required and the one we retain is that of viscosity solution. We provide it for the evolutionary equation even though it is not explicitly needed. It is left to the reader to infer the analogous definition for the stationary equation. A good introduction to the subject is [33]:

Definition 2.5.1

- A continuous function $u : [0, +\infty) \times M \to \mathbb{R}$ is a viscosity subsolution to (EHJ) if it verifies the initial condition and if for any C^1 function $\phi : (0, +\infty) \times M \to \mathbb{R}$, if $u - \phi$ has a local maximum at (t_0, x_0) then

$$\partial_t \phi(t_0, x_0) + H\big(x_0, \partial_x \phi(t_0, x_0)\big) \leqslant 0.$$

- A continuous function $u : [0, +\infty) \times M \to \mathbb{R}$ is a viscosity supersolution to (EHJ) if it verifies the initial condition and if for any C^1 function $\phi : (0, +\infty) \times M \to \mathbb{R}$, if $u - \phi$ has a local minimum at (t_0, x_0) then

$$\partial_t \phi(t_0, x_0) + H\big(x_0, \partial_x \phi(t_0, x_0)\big) \geqslant 0.$$

- A continuous function $u : [0, +\infty) \times M \to \mathbb{R}$ is a viscosity solution to (EHJ) if it is both a subsolution and a supersolution.

In the rest of the exposition, unless otherwise specified, any solution, subsolution or supersolution will be implicitly understood in the viscosity sense and the adjective will be omitted.

This notion is particularly adapted to our problem for several reasons. The first one is that it is reasonable in the sense that if a solution u is differentiable at $(t_0, x_0) \in (0, +\infty) \times M$ then it solves the Hamilton–Jacobi equation at that point:

$$\partial_t u(t_0, x_0) + H\big(x_0, \partial_x u(t_0, x_0)\big) = 0.$$

As it can be proved that in our setting, solutions are locally Lipschitz, viscosity solutions turn out to be almost everywhere solutions. However, beware that the converse is not true. The following theorem makes viscosity solutions particularly handy (see [33, 92]):

Theorem 2.5.2 *Given a continuous function $u_0 : M \to \mathbb{R}$, there exists a unique solution to* (EHJ). *This solution will be denoted $(x, t) \mapsto S^-(t)(u_0)(x)$.*

For any fixed $t > 0$, the operator $S^-(t)$ is acting on $C^0(M, \mathbb{R})$. Due to the uniqueness of solutions and to the fact that H is autonomous, S^- is a semigroup, meaning that $S^-(t+s) = S^-(t) \circ S^-(s)$. It turns out it enjoys properties very similar to the discrete Lax–Oleinik semigroup T^-:

Proposition 2.5.3

1. *For any $t > 0$, there exists $K > 0$ such that the set $S^-(t)\big(C^0(M, \mathbb{R})\big)$ contains only K-Lipschitz functions.*
2. *For any $t > 0$, $S^-(t)$ commutes with addition of constants: $S^-(t)(f + k) = S^-(t)(f) + k$, for all $f \in C^0(M, \mathbb{R})$ and $k \in \mathbb{R}$.*
3. *For any $t > 0$, $S^-(t)$ is order preserving: if $f \leqslant g$ then $S^-(t)f \leqslant S^-(t)g$.*
4. *For any $t > 0$, $S^-(t)$ is 1-Lipschitz for the sup-norm.*

At this stage, the similarities between discrete weak KAM theory and the Hamilton–Jacobi equations are not clear. It comes from an explicit control-theoretic representation formula of the operators $S^-(t)$. Let us define the Lagrangian as the convex dual of the Hamiltonian:

Definition 2.5.4 The Lagrangian $L : TM \to \mathbb{R}$ is defined by

$$\forall (x, v) \in TM, \quad L(x, v) = \sup_{p \in T_x^* M} p(v) - H(x, p).$$

In this definition, the supremum is actually a maximum. The Lagrangian L is termed *Tonelli Lagrangian* as it enjoys the same properties as H:

- L is C^2,
- L is strictly convex in the speed variable, meaning that for any $(x, v) \in TM$ the Hessian $\partial_{vv} L(x, v)$ is positive definite.
- L is superlinear, meaning that

$$\forall x \in M, \quad \lim_{\|v\|_x \to +\infty} \frac{L(x, v)}{\|v\|_x} = +\infty.$$

Through the Lagrangian and Hamiltonian functions we can go from TM to T^*M thanks to the Fenchel transform $\mathcal{L}$ defined by

$$\forall (x,v) \in TM, \quad \mathcal{L}(x,v) = \big(x, \partial_v L(x,v)\big) \in T^*M. \tag{2.2}$$

This transformation is a C^1 diffeomorphism under the Tonelli assumptions and its inverse is given by

$$\forall (x,p) \in T^*M, \quad \mathcal{L}^{-1}(x,p) = \big(x, \partial_p H(x,p)\big) \in TM.$$

Moreover, H and L are also related by the formulas

$$H\big(x, \partial_v L(x,v)\big) = \partial_v L(x,v)(v) - L(x,v) \ ;$$
$$L\big(x, \partial_p H(x,p)\big) = p\big(\partial_p H(x,p)\big) - H(x,p).$$

Theorem 2.5.5 *Let $u : M \to \mathbb{R}$ be any continuous function. For any $t > 0$ and $x \in M$ the following holds:*

$$S^-(t)u(x) = \inf_{\substack{\gamma:[-t,0]\to M \\ \gamma(0)=x}} u\big(\gamma(-t)\big) + \int_{-t}^{0} L\big(\gamma(s), \dot\gamma(s)\big)\mathrm{d}s. \tag{2.3}$$

In this formula, called the *Lax–Oleinik formula*, the infimum is taken amongst absolutely continuous curves. Tonelli theory asserts that the infimum is a minimum and any such minimum turns out to be C^2 and to verify the Euler–Lagrange equation: $\frac{\mathrm{d}}{\mathrm{d}t}\partial_v L(\gamma,\dot\gamma) = \partial_x L(\gamma,\dot\gamma)$ (see [81]). This equation defines a complete flow on TM denoted by φ_L. It is called the *Euler–Lagrange flow*. Its conjugate by the Fenchel transform $\varphi_H = \mathcal{L} \circ \varphi_L \circ \mathcal{L}^{-1}$ is a flow on T^*M called *Hamiltonian flow*. Its trajectories solve Hamilton's equations:

$$\begin{cases} \dot x = \partial_p H(x,p), \\ \dot p = -\partial_x H(x,p). \end{cases} \tag{2.4}$$

The infimum in (2.3) can be split in two by first choosing a starting point y for the curves and then minimizing between y and x. More precisely, if we define the action functional

$$\forall (t,y,x) \in [0,+\infty) \times M \times M, \quad h_t(y,x) = \inf_{\substack{\gamma:[-t,0]\to M \\ \gamma(0)=x \\ \gamma(-t)=y}} \int_{-t}^{0} L\big(\gamma(s), \dot\gamma(s)\big)\mathrm{d}s, \tag{2.5}$$

then the formula for solutions of (EHJ) becomes

$$S^{-}(t)u(x) = \inf_{y \in M} u(y) + h_t(y, x),$$

which is exactly the discrete Lax–Oleinik semigroup with cost function $c = h_t$.

2.5.2 The Weak KAM Theorem and Critical Subsolutions

Another important fact follows from the simple remark that a function $u : M \to \mathbb{R}$ is solution to the stationary Hamilton–Jacobi equation (SHJ) with constant α if and only if the function $U(t, x) = u(x) - \alpha t$ is solution to the evolutionary Eq. (EHJ) with initial condition $u_0 = u$. Hence any such solution is characterized by the property

$$\begin{aligned}\forall (t, x) \in (0, +\infty) \times M, \quad u(x) &= \inf_{\substack{\gamma:[-t,0]\to M \\ \gamma(0)=x}} u\big(\gamma(-t)\big) \\ &\quad + \int_{-t}^{0} \Big[L\big(\gamma(s), \dot\gamma(s)\big) + \alpha\Big] \mathrm{d}s \\ &= \inf_{y \in M} u(y) + h_t(y, x) + t\alpha.\end{aligned}$$

Note also that such solutions verify in particular that[4]

$$\begin{aligned}&\forall t > 0,\ \forall \gamma : [-t, 0] \to M, \quad u\big(\gamma(0)\big) - u\big(\gamma(-t)\big) \\ &\leqslant \int_{-t}^{0} \Big[L\big(\gamma(s), \dot\gamma(s)\big) + \alpha\Big] \mathrm{d}s. \qquad (2.6)\end{aligned}$$

In fact, verifying the above family of inequalities characterizes u to be a subsolution of (SHJ).

With these facts in mind, it should not come as a surprise that the original weak KAM Theorem of Fathi [105] is similar to the discrete weak KAM Theorem we stated:

Theorem 2.5.6 *There exists a unique constant $\alpha(0) \in \mathbb{R}$ for which the stationary Eq.* (EHJ) *admits a solution with right hand side equal to $\alpha = \alpha(0)$.*[5]

---

[4] As already implied, when developing the theory, the curves considered are at least absolutely continuous. A posteriori, all interesting minimizing curves are in fact of class C^2. The reader can therefore think of all curves as of C^2 curves.

[5] The notation $\alpha(0)$ is borrowed from Mather's α function. It is a function acting on the first cohomology group of M. Given a closed 1-form c, one can perturb the stationary Eq. (SHJ) by

Proof For fun's sake, we provide yet another proof of this Theorem. We deduce it from the discrete weak KAM theorem, although it is highly unnatural.

Uniqueness of $\alpha(0)$ follows from the uniqueness of the critical constant in the discrete weak KAM Theorem 2.2.1 as a classical weak KAM solution is also a discrete weak KAM solution for the cost h_1 and same critical constant.

The first (and central) part of the proof is to establish that for $t > 0$ fixed, h_t is Lipschitz continuous (hence continuous). Note that the second point of Proposition 2.5.3 also follows from that assertion. We omit this technical (and central) aspect and refer to [109].

We now apply the discrete weak KAM Theorem 2.2.1 which states that for all $n \in \mathbb{N}$, there exists a unique constant c_n and a function $u_n : M \to \mathbb{R}$ such that $u_n = S^-(2^{-n})u_n + c_n$. By using the semigroup property. One obtains that

$$u_n = \left(S^-(2^{-n})\right)^{2^n} u_n + 2^n c_n = S^-(1)u_n + 2^n c_n.$$

It follows by the uniqueness of c_0 that for all $n \geqslant 0$, $c_n = 2^{-n}c_0$. The same argument yields that

$$\forall n \geqslant 0,\ \forall t \in 2^{-n}\mathbb{N}, \quad u_n = S^-(t)u_n + tc_0. \tag{2.7}$$

Up to adding constants to the functions u_n we may assume they all vanish at some point of M which does not change their property of being weak KAM solutions for $S^-(2^{-n})$. Moreover, as all the $u_n, n \in \mathbb{N}$ are in the image of $S^-(1)$ they form an equi-Lipschitz family of functions. Hence by the Arzelà–Ascoli theorem, we may find an extracted sequence k_n such that $(u_{k_n})_{n\in\mathbb{N}}$ converges uniformly to a function v. By continuity of $u \mapsto S^-(t)u$ we may pass to the limit (as $m \to +\infty$) in the equalities

$$\forall m \geqslant n,\ \forall t \in 2^{-k_n}\mathbb{N}, \quad u_{k_m} = S^-(t)u_{k_m} + tc_0,$$

to obtain that $v = S^-(t)v + tc_0$ for any diadic number t. As diadic numbers are dense in $[0, +\infty)$ the theorem follows again by continuity of $t \mapsto S^-(t)v$. □

We have used at the end of the proof the following result, of which we give a quick proof for completeness:

Lemma 2.5.7 *Let $u : M \to \mathbb{R}$ be a continuous function, then the function $t \mapsto S^-(t)u$ is uniformly continuous.*

Proof Using the semigroup property and non-expansiveness, one obtains that for all $0 \leqslant s \leqslant t$, $\|S^-(t)u - S^-(s)u\|_\infty \leqslant \|S^-(t-s)u - u\|_\infty$. Therefore, it is enough to prove continuity at 0. Moreover, again using the non-expansive character, one

$H(x, c(x) + D_x u) = \alpha$ and prove a weak KAM theorem for this equation. The critical constant found depends only on the cohomology class $[c]$ and is $\alpha([c])$. Discrete analogues of this are discussed in [201].

sees by an approximation argument that it is enough to prove the result for u being a Lipschitz function.

Assume therefore that u is Lipschitz continuous with Lipschitz constant $K > 0$. As L is superlinear, there is a constant $C > 0$ such that

$$\forall (x,v) \in TM, \quad L(x,v) \geqslant K\|v\|_x - C.$$

It follows that for any absolutely continuous $\gamma : [-t, 0] \to M$,

$$\int_{-t}^{0} L\big(\gamma(s), \dot\gamma(s)\big)\mathrm{d}s \geqslant \int_{-t}^{0} K\|\dot\gamma(s)\|_{\gamma(s)}\mathrm{d}s - tC \geqslant K\mathrm{d}\big(\gamma(0), \gamma(t)\big) - tC.$$

It follows that if $\gamma(0) = x$, recalling that u is K-Lipschitz,

$$\int_{-t}^{0} \Big[L\big(\gamma(s), \dot\gamma(s)\big) + \alpha\Big]\mathrm{d}s + u\big(\gamma(-t)\big) - u(x) \geqslant$$
$$\geqslant K\mathrm{d}\big(\gamma(0), \gamma(t)\big) - tC - K\mathrm{d}\big(\gamma(0), \gamma(t)\big)$$

and taking an infimum on all curves, $S^-(t)u(x) - u(x) \geqslant -tC$.

Finally, comparing with a constant curve in the definition of the Lax–Oleinik semigroup one finds that

$$S^-(t)u(x) \leqslant u(x) + \int_{-t}^{0} L(x,0)\mathrm{d}s \leqslant u(x) + t\max_{y\in M} L(y,0).$$

Those two inequalities prove the lemma. □

One may wonder if there is a relationship between discrete weak KAM solutions and weak KAM solutions. It turns out that the answer is yes and it is closely related to the autonomous aspect of our setting.[6]

Theorem 2.5.8 *Let $c = h_1$ be the cost function. Then we have $\alpha(0) = c[0]$. Moreover a function u is a discrete weak KAM solution for c if and only if it is a weak KAM solution for H.*

The proof heavily relies on Fathi's Theorem [107]:

Theorem 2.5.9 (Fathi [107]) *Let $v : M \to \mathbb{R}$ be a continuous function. Then $t \mapsto S^-(t)v + t\alpha(0)$ uniformly converges to a weak KAM solution, for H, as $t \to +\infty$.*

Proof of Theorem 2.5.8 It is clear that if u is a weak KAM solution, then $u = S^-(1)u + \alpha(0) = T^-u + \alpha(0)$. It follows that $\alpha(0) = c[0]$ because of the uniqueness

[6] The second part of the following Theorem becomes tautological when considering time-dependent 1-periodic Hamiltonians as the definitions of weak KAM solutions then coincide.

in Theorem 2.2.1. Moreover, any weak KAM solution is a discrete weak KAM solution.

Let now v be a discrete weak KAM solution. It follows that $v = S^-(1)v + \alpha(0)$ and then that

$$\forall n \in \mathbb{N}, \quad v = S^-(n)v + n\alpha(0).$$

By Fathi's theorem, there exists a weak KAM solution $\tilde{v}$ such that $S^-(t)v + t\alpha(0) \to \tilde{v}$ as $t \to +\infty$. It then follows that $v = \tilde{v}$ is a weak KAM solution. □

If $\alpha \in \mathbb{R}$, denote by $\mathcal{S}'_\alpha$ the set of subsolutions to (SHJ), or equivalently functions verifying (2.6). Note that $u \in \mathcal{S}'_\alpha$ if and only if $t \mapsto S^-(t)u + t\alpha$ is non-decreasing. Finally, $\mathcal{S}'$ will denote the special set of critical subsolutions $\mathcal{S}'_{\alpha(0)}$. As will be seen later, unlike what happens for weak KAM solutions, the set $\mathcal{S}'_\alpha \subset \mathcal{S}_\alpha$ is much smaller than its discrete analogue. The first point of the next proposition is a hint as to why. We state without proof analogous results to Proposition 2.2.6:

Proposition 2.5.10 *Let $\alpha \in \mathbb{R}$, the following assertions hold:*

(i) The family of α-subsolutions is equi-Lipschitz.
(ii) The set $\mathcal{S}'_\alpha$ of α-subsolutions is closed (with respect to pointwise and uniform convergence), convex and stable by the Lax–Oleinik semigroup: $S^-(t)(\mathcal{S}'_\alpha) \subset \mathcal{S}'_\alpha$ for all $t \geqslant 0$.

And here is the characterization of $\alpha(0)$ similar to Proposition 2.2.7:

Proposition 2.5.11 *The following equality holds:* $\alpha(0) = \min\{\alpha \in \mathbb{R},\ \ \mathcal{S}'_\alpha \neq \varnothing\}$.

We finish this section by discussing the continuous way of adapting the second proof of the discrete weak KAM theorem. Indeed, if $\lambda \in (0, 1)$, considering the mapping $u \mapsto S^-(1)(\lambda u)$ does not make much sense from the PDE point of view.

The function u_λ constructed in the second proof of Theorem 2.2.1 satisfies $T^-(\lambda u_\lambda) = u_\lambda$. By setting $v_\lambda = \lambda u_\lambda$, the previous equation may be rewritten as

$$v_\lambda - T^-(v_\lambda) = (\lambda - 1)u_\lambda = (1 - \lambda^{-1})v_\lambda.$$

We now follow the intuition that T^-stands for the time 1 of an evolution semigroup, hence $v_\lambda - T^-(v_\lambda)$ can be interpreted as a discrete derivative with respect to time.

Going back to the Hamilton–Jacobi equation, and following the previous analysis, we are looking for a function $u : M \to \mathbb{R}$ such that

$$\frac{\mathrm{d}}{\mathrm{d}t} S^-(t)u_{|t=0} = -\ell u,$$

where we applied the change of variable $\ell = \lambda^{-1} - 1$. Hence $\ell > 0$ is aimed to go to 0. Remembering that $(t, x) \mapsto S^-(t)u(x)$ solves the Hamilton–Jacobi equation, we are actually looking for a function u solving

$$\ell u(x) + H(x, D_x u) = 0, \quad x \in M \tag{ℓHJ}$$

in the viscosity sense. The preceding equation is called the *discounted equation* and ℓ is called the *discount factor*. It turns out this is precisely the method used in [152] which is historically the first paper where weak KAM solutions appear. In this foundational preprint, they prove the following results, for a wider class of Hamiltonians (in particular, no convexity is required):

Theorem 2.5.12

1. *For all $\ell > 0$ there exists a unique function $U_\ell : M \to \mathbb{R}$ which is a viscosity solution of* (ℓHJ).
2. *The family $(\ell U_\ell)_{\ell>0}$ is equi-bounded.*
3. *The family $(U_\ell)_{\ell>0}$ is equi-Lipschitz.*
4. *Given $x_0 \in M$ and setting $v_\ell = U_\ell - U_\ell(x_0)$, it follows that the family $(v_\ell)_{\ell>0}$ is relatively compact.*
5. *The family $(\ell U_\ell)_{\ell>0}$ uniformly converges to the constant function $-\alpha(0)$[7] as $\ell \to 0$ and any limit function v_0 of the family $(v_\ell)_{\ell>0}$, as $\ell \to 0$, is a solution of* (SHJ) *with right hand side $\alpha(0)$.*

2.5.3 The Positive Lax–Oleinik Semigroup

As in our discrete setting, Fathi introduced the positive Lax–Oleinik semigroup as follows:

Definition 2.5.13 Let $u : M \to \mathbb{R}$ be any continuous function, for any $t > 0$ and $x \in M$ we define:

$$S^+(t)u(x) = \sup_{\substack{\gamma:[0,t]\to M \\ \gamma(0)=x}} u\big(\gamma(t)\big) - \int_0^t L\big(\gamma(s), \dot\gamma(s)\big)\mathrm{d}s = \sup_{y\in M} u(y) - h_t(x, y),$$

where the supremum is taken amongst absolutely continuous curves.

[7] In the preprint, this constant is actually denoted by $-\overline{H}(0)$ and $\overline{H}$ is called the *effective Hamiltonian.*

Once again, this semigroup has a natural PDE interpretation. Let us introduce the Hamiltonian: $\check{H} : (x, p) \mapsto H(x, -p)$. One verifies that the associated Lagrangian is given by

$$\forall (x, v) \in TM, \quad \check{L}(x, v) = \sup_{p \in T_x^* M} p(v) - \check{H}(x, p) = \sup_{p \in T_x^* M} p(-v) - H(x, p)$$
$$= L(x, -v).$$

Therefore the positive semigroup is written as follows:

$$S^+(t)u(x) = - \inf_{\substack{\gamma:[0,t]\to M \\ \gamma(0)=x}} -u\big(\gamma(t)\big) + \int_0^t L\big(\gamma(s), \dot\gamma(s)\big)\mathrm{d}s$$
$$= -\left[\inf_{\substack{\check\gamma:[-t,0]\to M \\ \check\gamma(0)=x}} -u\big(\check\gamma(t)\big) + \int_{-t}^0 \check{L}\big(\check\gamma(s), \dot{\check\gamma}(s)\big)\mathrm{d}s \right]$$
$$= -\check{S}(t)(-u)(x),$$

where $\check{S}$ denotes the Lax–Oleinik semigroup associated to $\check{H}$. Hence the function $(t, x) \mapsto -S^+(t)u(x)$ solves an evolutionary Hamilton–Jacobi equation with Hamiltonian $\check{H}$ and initial data $-u$.

As the Hamiltonian $\check{H}$ is also Tonelli, it is automatic that the positive semigroup S^+ has the same properties as S^- stated in Proposition 2.5.3, that we do not recall here. As for the discrete case, a positive weak KAM theorem follows:

Theorem 2.5.14 (Positive Weak KAM) *The critical constant $\alpha(0)$ is the only one for which the equation $u = S^+(t)u - t\alpha(0)$, for all $t > 0$, admits solutions $u : M \to \mathbb{R}$. Moreover the constant $\alpha(0)$ has the following characterization:*

$$\forall v \in \mathcal{B}(M, \mathbb{R}), \quad \frac{S^+(t)v}{t} \underset{t\to+\infty}{\longrightarrow} \alpha(0),$$

and the convergence is uniform.

2.5.4 Strict Subsolutions, Aubry Sets

As is expected, given a constant $\alpha \in \mathbb{R}$, a function $u : M \to \mathbb{R}$ is an α-subsolution ($u \in \mathcal{S}'_\alpha$) if

$$\forall t > 0,\ \forall \gamma : [-t, 0] \to M, \quad u\big(\gamma(0)\big) - u\big(\gamma(-t)\big) \leqslant \int_{-t}^0 \Big[L\big(\gamma(s), \dot\gamma(s)\big) + \alpha\Big]\mathrm{d}s,$$

where γ ranges in the set of absolutely continuous curves. It can be established that subsolutions are automatically Lipschitz hence differentiable almost everywhere.

The terminology comes from the fact that subsolutions are indeed viscosity subsolutions to the critical stationary Hamilton–Jacobi equation (SHJ). Whence the following characterizations of subsolutions hold:

Proposition 2.5.15 *Let $u : M \to \mathbb{R}$ and $\alpha \in \mathbb{R}$ be a constant. The following are equivalent:*

1. *$u \in \mathcal{S}'_\alpha$ is an α-subsolution;*
2. *the family of functions $\big(S^-(t)u + t\alpha\big)_{t \geqslant 0}$ is non-decreasing;*
3. *the family of functions $\big(S^+(t)u - t\alpha\big)_{t \geqslant 0}$ is non-increasing;*
4. *the function u verifies $H(x, D_x u) \leqslant \alpha$ in the viscosity sense;*
5. *the function u verifies $H(x, D_x u) \leqslant \alpha$ for almost every $x \in M$ (more precisely at every $x \in M$ such that u is differentiable at x).*

We now focus on the particular case of critical subsolutions. The idea is that, because the constant $\alpha(0)$ is a threshold between a world with subsolutions and a world without, there is no critical subsolution where the inequality $H(x, D_x u) \leqslant \alpha(0)$ is everywhere strict. More precisely, the obstruction to having strict inequalities is concentrated on a subset of M: the projected Aubry set. This set was introduced for twist maps (that are a particular discrete setting) by Aubry and Le Daeron [27], Mather [159], and Mather and Forni [172]. For Hamiltonian systems, it was later on introduced by Mather [165] by dynamical systems means. The present interpretation, using critical subsolutions, is due to Fathi [104–106, 109].

A fundamental result on subsolutions is due to Fathi and Siconolfi [114, 115] (for C^1 subsolutions) and was then improved by Bernard [42]:

Theorem 2.5.16 *There exists a function $u_0 : M \to \mathbb{R}$ that is $C^{1,1}$ (C^1 with Lipschitz differential) and is a critical subsolution ($u_0 \in \mathcal{S}'_{\alpha_0}$). Moreover it verifies the following fundamental property:*

if for some $x \in M$, $H(x, D_x u_0) = \alpha(0)$ then if $u \in \mathcal{S}'_{\alpha(0)}$ is any other critical subsolution, u is differentiable at x and $D_x u = D_x u_0$. Hence $H(x, D_x u) = \alpha(0)$.

The function u_0 above is a $C^{1,1}$ strict subsolution. The set

$$\mathcal{A} = \{x \in M, \quad H(x, D_x u_0) = \alpha(0)\},$$

is called the *projected Aubry set*. It is non-empty $\big($otherwise u_0 would be a subsolution for a constant less that $\alpha(0)\big)$. It is also compact. The Aubry set is $\mathcal{A}^* = \{(x, D_x u_0), \quad x \in \mathcal{A}\} \subset T^*M$. Of course, $\mathcal{A}^* \subset H^{-1}(\{\alpha(0)\})$. This is actually a deep fact. It implies a Theorem of Carneiro [74] which we come back to later. It may seem at first glance that those sets depend on u_0 but it is not the case. Indeed, $\mathcal{A}$ is a set of points where all subsolutions u are differentiable[8] and $H(x, D_x u) = \alpha(0)$. The Aubry set in TM is then naturally $\mathcal{A}' = \mathcal{L}^{-1}(\mathcal{A}^*)$.

As we saw in the discrete setting, points of the projected Aubry set come in whole sequences, giving rise to the Aubry set (subset of $X^{\mathbb{Z}}$). Moreover, this Aubry set is invariant by the shift on $X^{\mathbb{Z}}$. Similar phenomena arise in the classical setting, the dynamical systems here being those of φ_L the Euler–Lagrange flow and φ_H the Hamiltonian flow:

Theorem 2.5.17 *The Aubry set $\mathcal{A}^*$ is invariant by φ_H meaning that for all $t \in \mathbb{R}$, $\varphi_H^t(\mathcal{A}^*) = \mathcal{A}^*$. Equivalently, the Aubry set $\mathcal{A}'$ is invariant by φ_L meaning that for all $t \in \mathbb{R}$, $\varphi_L^t(\mathcal{A}') = \mathcal{A}'$.*

It follows that if $(x, p) \in \mathcal{A}^*$, $u \in \mathcal{S}'_{\alpha(0)}$ is a critical subsolution ant $t \in \mathbb{R}$, then $\big(x(t), p(t)\big) = \big(x(t), D_{x(t)}u\big)$ where we set $\big(x(t), p(t)\big) = \varphi_H^t(x, p)$. In particular, we stress that u is automatically differentiable at $x(t)$ for all $t \in \mathbb{R}$.

The curve $\big(x(t), \dot{x}(t)\big)_{t\in\mathbb{R}}$ is a trajectory of the Euler–Lagrange flow. Moreover, its projection on M calibrates u in the sense that

$$\forall s < t, \quad u\big(x(t)\big) - u\big(x(s)\big) = \int_s^t L\big(x(\sigma), \dot{x}(\sigma)\big)\mathrm{d}\sigma + (t - s)\alpha(0).$$

In particular, recalling that u verifies (2.6), *it follows that*

$$\begin{aligned}\forall t > 0, \quad S^-(t)u(x) + t\alpha(0) &= u\big(x(-t)\big) + \int_{-t}^0 L\big(x(\sigma), \dot{x}(\sigma)\big)\mathrm{d}\sigma \\ &\quad + t\alpha(0) = u(x); \\ S^+(t)u(x) - t\alpha(0) &= u\big(x(t)\big) - \int_0^t L\big(x(\sigma), \dot{x}(\sigma)\big)\mathrm{d}\sigma \\ &\quad - t\alpha(0) = u(x).\end{aligned}$$

From this follows the analogue of Corollary 2.4.10:

Corollary 2.5.18 *Let $x \in M$. The following assertions are equivalent:*

(i) $x \in \mathcal{A}$,
(ii) $\forall u \in \mathcal{S}'_{\alpha(0)}$, $\forall t > 0$, $\quad u(x) = S^-(t)u(x) + t\alpha(0)$,
(iii) $\forall u \in \mathcal{S}'_{\alpha(0)}$, $\forall t > 0$, $\quad u(x) = S^+(t)u(x) - t\alpha(0)$.

We end this section with a new description of the relation between the Aubry set for the Hamiltonian and the Aubry set for the time-1 action functional h_1. For the sake of clarity, we denote them respectively $\mathcal{A}_H$ and $\mathcal{A}_{h_1}$.

Theorem 2.5.19 *The equality $\mathcal{A}_H = \mathcal{A}_{h_1}$ holds.*

[8] This actually suffices to characterize the projected Aubry set as proven in [114].

Proof Recall that if $n > 0$, thanks to the choice of cost function, the Lax–Oleinik semigroups are linked by the equalities $T^{-n} = S^-(n)$ and $T^{+n} = S^+(n)$. Moreover, we have seen that both critical constants for H and h_1 coincide. It follows that classical subsolutions are discrete subsolutions, $\mathcal{S}'_{\alpha(0)} \subset \mathcal{S}_{\alpha(0)}$ (the inclusion being very often strict). Indeed, $u \in \mathcal{S}'_{\alpha(0)}$ if and only if $t \mapsto S^-(t)u + t\alpha(0)$ is non-decreasing for $t \in [0, +\infty)$ while $u \in \mathcal{S}_{\alpha(0)}$ if and only if the sequence $n \mapsto S^-(n)u + n\alpha(0)$ is non-decreasing for $n \geqslant 0$.

Let now $x \in \mathcal{A}_{h_1}$ and $u \in \mathcal{S}'_{\alpha(0)}$. We deduce from the preceding discussion and Corollary 2.4.10 that the sequence $n \mapsto S^-(n)u(x) + n\alpha(0)$ is constant. As the family $t \mapsto S^-(t)u + t\alpha(0)$ is non-decreasing, it has to be constant. This being true for any classical subsolution, we deduce that $x \in \mathcal{A}_H$ by Corollary 2.5.18. We have established that $\mathcal{A}_{h_1} \subset \mathcal{A}_H$.

Let then $x \in \mathcal{A}_H$ and $u \in \mathcal{S}_{\alpha(0)}$ be a discrete subsolution. We set $u_- = \lim_{n\to+\infty} S^-(n)u + n\alpha(0)$ and $u_+ = \lim_{n\to+\infty} S^+(n)u - n\alpha(0)$ that both exist (the sequences are monotonous). They are respectively a negative and positive weak KAM solution and verify $u_+ \leqslant u \leqslant u_-$. Finally, let us set $v = \lim_{t\to+\infty} S^-(t)u_+ + t\alpha(0)$ that is a negative weak KAM solution. As classical and discrete weak KAM solutions coincide, we know from Corollary 2.5.18 that $u_+(x) = S^-(t)u_+(x) + t\alpha(0)$ for all $t > 0$, hence $u_+(x) = v(x)$.

We will prove that $v = u_-$. Let $\varepsilon > 0$ and $N > 0$ such that $\|S^+(n)u - n\alpha(0) - u_+\| < \varepsilon$ for all $n \geqslant N$. By application of Proposition 2.3.6 and monotonicity of the semigroups, one establishes that $S^-(n) \circ S^+(n)u \geqslant u$. By taking $n > N$ and using the monotonicity of S^- again, it follows that

$$S^-(n)u_+ + n\alpha(0) \geqslant S^-(n) \circ S^+(n)u - \varepsilon \geqslant u - \varepsilon.$$

Then if $m > 0$ we obtain that $S^-(n+m)u_+ + (n+m)\alpha(0) \geqslant S^-(m)u + m\alpha(0) - \varepsilon$. Finally letting $m \to +\infty$ it follows that $v \geqslant u_- - \varepsilon$, and as this is true for all $\varepsilon > 0$ we obtain that $v \geqslant u_-$.

The reverse inequality is easier: as $u_+ \leqslant u_-$ then

$$\forall n > 0, \quad S^-(n)u_+ + n\alpha(0) \leqslant S^-(n)u_- + n\alpha(0) = u_-,$$

and we conclude by letting $n \to +\infty$.

Finally, it follows that $u_+(x) = v(x) = u_-(x)$ and as for $n > 0$,

$$u_+ \leqslant u \leqslant S^-(n)u + n\alpha(0) \leqslant u_-,$$

we also have $u(x) = S^-(n)u(x) + n\alpha(0) = u_-(x)$. Thus $x \in \mathcal{A}_{h_1}$ and the proof is complete. □

In the preceding proof the functions u_- and u_+ form what is called a *conjugate pair* (see [104] for the classical definition and [49] for definitions in a discrete setting), a notion that will reappear in this text (as in Remark 3.1.4).

Chapter 3
More (Dynamical) Characterizations of the Aubry Sets

Abstract This Chapter provides further properties and characterizations of the various Aubry sets introduced in the previous chapter. We provide examples of particular points in the Aubry set. Finally, by adding suitable twist hypotheses, we recover classical graph properties of the Aubry sets. We conclude by showing how costs coming from Tonelli Hamiltonians satisfy those twist conditions.

So far, we have constructed several different Aubry sets and have started to study the way they are related. Even though their initial definition stems from one particular subsolution, we already began to understand that, in the end, they entail informations concerning all subsolutions. In fact they depend on the cost function c only, and not on the particular subsolution u_0. In this section we push further the understanding of the meaning of those Aubry sets.

Most results of this Chapter are to be found in [202] by the author for the most abstract and general part. When it comes to the more regular setting of costs defined on a compact manifold first results are written in [201] by the author, then generalized in [54] with Bernard.

3.1 The Peierls Barrier

The aim of this paragraph is to give a characterization of the Aubry set in terms of action along long chains of points. This characterization builds on the following notion:

Definition 3.1.1 If $n > 0$ is a positive integer and $(x, y) \in X \times X$, let

$$c_n(x, y) = \inf \left\{ \sum_{i=0}^{n-1} c(x_i, x_{i+1}),\ (x_0, x_1, \cdots, x_{n-1}, x_n) \in X^{n+1},\ x_0 = x,\ x_n = y \right\}.$$

M. Zavidovique, *Discrete Weak KAM Theory*, Lecture Notes in Mathematics 2377,
https://doi.org/10.1007/978-3-031-96809-9_3

The *Peierls barrier* is the function $h : X \times X \to X$ defined by

$$\forall (x, y) \in X \times X, \quad h(x, y) = \liminf_{n \to +\infty} c_n(x, y) + nc[0].$$

It follows from the previous Definition that if $u \in \mathcal{B}(X, \mathbb{R})$ is a bounded function, then $T^{-n}u(x) = \inf_{y \in X} u(y) + c_n(y, x)$.

The name Peierls barrier appears in Aubry and Le Daeron's work [27] in the context of Conservative Twist Maps of the Annulus (see the last Chapter of this text). It associates to each rotation number a real number that vanishes if and only if there is an invariant circle with this rotation number (all those notions are defined in the last Chapter). Hence it is a barrier to the existence of invariant circles. Mather then introduced variations and studied properties of this barrier, still in the context of Twist Maps in [161]. He went on to generalize the Peierls barrier to higher dimensional Lagrangian settings in [167]. The Definition just presented is analogous to what is done in the latter reference.

Let us start by giving some properties of this new object:

Proposition 3.1.2

(i) *h is well defined and continuous;*
(ii) *for all $u \in \mathcal{S}$ and $(x, y) \in X \times X$, $u(y) - u(x) \leqslant h(x, y)$, in particular for all $x \in X$, $h(x, x) \geqslant 0$;*
(iii) *for all $x, y, z \in X$ and integer $n > 0$ the following inequalities hold:*

$$\begin{aligned} h(x, y) &\leqslant h(x, z) + c_n(z, y) + nc[0]; \\ h(x, y) &\leqslant c_n(x, z) + nc[0] + h(z, y); \\ h(x, y) &\leqslant h(x, z) + h(z, y); \end{aligned} \tag{3.1}$$

(iv) *for all $x \in X$, the function $h_x = h(x, \cdot)$ is a weak KAM solution and the function $h^x = -h(\cdot, x)$ is a positive weak KAM solution.*

Proof Let ω denote a modulus of continuity for c: a non-decreasing function from $[0, +\infty)$ to itself that is continuous at 0, with $\omega(0) = 0$, such that

$$\forall (x, y, x', y') \in X^4, \quad |c(x, y) - c(x', y')| \leqslant \omega\big(d(x, x') + d(y, y')\big).$$

Let x, x', y, y' be points, n an integer, and $x = x_0, x_1, \cdots, x_{n-1}, x_n = y$ such that $c_n(x, y) = \sum_{i=0}^{n-1} c(x_i, x_{i+1})$ (they exist by compactness of X and continuity of the maps). Then one gets by definition that

$$\begin{aligned} c_n(x', y') - c_n(x, y) &\leqslant c(x', x_1) + \sum_{i=1}^{n-2} c(x_i, x_{i+1}) + c(x_{n-1}, y') - \sum_{i=0}^{n-1} c(x_i, x_{i+1}) \\ &\leqslant 2\omega\big(d(x, x') + d(y, y')\big). \end{aligned}$$

We get the same inequality for $c_n(x, y) - c_n(x', y')$ by the same argument, which proves that the c_n are equicontinuous. Moreover, one checks readily that $c_{n+1}(x, y) = T^{-n}c(x, \cdot)(y)$. Therefore, combined with the first point of Remark 2.2.4, this implies that the family on functions $c_n + nc[0]$ is uniformly bounded. This proves (i).

We have already seen in Remark 2.4.3-(iii) that if $u \in \mathcal{S}$ and x, y are points, then for any integer $n > 0$, $u(y) - u(x) \leqslant c_n(x, y) + nc[0]$. Taking a liminf yields (ii).

One has by definition that if m, n are integers and x, y are points, then

$$c_{n+m}(x, y) = \inf_{Z \in X} c_m(x, Z) + c_n(Z, y).$$

Hence if z is a third point, then

$$c_{n+m}(x, y) + (n + m)c[0] \leqslant c_m(x, z) + mc[0] + c_n(z, y) + nc[0].$$

Letting $m \to +\infty$ and taking liminf leads to the first inequality of (iii), $n \to +\infty$ and taking liminf to the second one, and both at the same time to the third.

Let $x \in X$. The previous inequalities applied to $n = 1$ show that the functions h_x and h^x are subsolutions. Let us prove (iv) for h_x, the rest being similar. Let $y \in X$ and k_n be an extraction such that $h_x(y) = \lim_{n\to+\infty} c_{k_n+1}(x, y) + (k_n + 1)c[0]$. For each n, there exists $x_n \in X$ such that $c_{k_n+1}(x, y) + (k_n + 1)c[0] = c_{k_n}(x, x_n) + c(x_n, y) + (k_n + 1)c[0]$ and taking a further extraction, one may assume that $x_n \to \tilde{x}$ for some $\tilde{x} \in X$. Taking the limit and by equicontinuity, one concludes that

$$h_x(y) = \lim_{n\to+\infty} c_{k_n}(x, x_n) + k_n c[0] + c(x_n, y) + c[0] \geqslant h(x, \tilde{x}) + c(\tilde{x}, y) + c[0].$$

The function h_x being a subsolution, the last inequality must be an equality and it follows that $h_x(y) = T^- h_x(y) + c[0]$ which is what was to be proven. □

We can actually strengthen (ii) in the previous Proposition as follows:

Proposition 3.1.3 *Let $u \in \mathcal{S}$ and m, n two non-negative integers, then*

$$\forall (x, y) \in X \times X, \quad T^{-n}u(y) - T^{+m}u(x) + (n + m)c[0] \leqslant h(x, y).$$

Proof Let m, n be any non-negative integers and $x = x_{-m}, \cdots, x_n = y$ be a chain of points. Then by definition of the Lax–Oleinik semigroups, one gets

$$\begin{aligned} T^{-n}u(y) - T^{+m}u(x) &\leqslant \sum_{i=0}^{n-1} c(x_i, x_{i+1}) + u(x_0) - u(x_0) + \sum_{i=-m}^{-1} c(x_i, x_{i+1}) \\ &= \sum_{i=-m}^{n-1} c(x_i, x_{i+1}). \end{aligned}$$

This being true for all chains of points it follows that

$$T^{-n}u(y) - T^{+m}u(x) + (n+m)c[0] \leqslant c_{n+m}(x, y) + (n+m)c[0].$$

Note that the left hand side is now non-decreasing in both n and m (as $u \in \mathcal{S}$). Therefore to show the statement, we only have to show that the limit when $n, m \to +\infty$ verifies the same inequality. Let n_k, m_k be two diverging increasing sequences such that $c_{n_k+m_k}(x, y) + (n_k + m_k)c[0] \to h(x, y)$. Then one concludes that

$$\begin{aligned} T^{-n}u(y) - T^{+m}u(x) + (n+m)c[0] &\leqslant \\ &\leqslant \lim_{k\to+\infty} T^{-n_k}u(y) - T^{+m_k}u(x) + (n_k+m_k)c[0] \\ &\leqslant \lim_{k\to+\infty} c_{n_k+m_k}(x, y) + (n_k+m_k)c[0] = h(x, y), \end{aligned}$$

which is the result. □

Remark 3.1.4 The previous proposition can actually be stated in a more optimal way by introducing another notion. Given a subsolution $u \in \mathcal{S}$, as noted, the sequences $T^{-n}u + nc[0]$ and $T^{+n}u - nc[0]$ converge respectively to functions u^- and u^+ which are respectively a negative and a positive weak KAM solution. The result then states that $u^-(y) - u^+(x) \leqslant h(x, y)$. Such pairs are called conjugate pairs, they coincide on $\mathcal{A}$ (Corollary 2.4.10) and this last point characterizes the pair (u^-, u^+) (see Proposition 3.2.5 in the next paragraph).

We are now ready to establish a characterization of the Aubry sets much easier to handle:

Theorem 3.1.5 *The following equalities hold:*

$$\mathcal{A} = \{x \in X, \quad h(x, x) = 0\},$$

$$\widehat{\mathcal{A}} = \{(x, y) \in X \times X, \quad c(x, y) + c[0] + h(y, x) = 0\}.$$

Beginning of the Proof We will only prove two inclusions for now. The other ones will arrive after an intermediate Proposition 3.1.6.

Let us call $A_1 \subset X$, $A_2 \subset X \times X$ the sets appearing in the right hand side of the statement. If $u \in \mathcal{S}$ and $h(x, x) = 0$, by the previous Proposition both sequences $(T^{-n}u(x)+nc[0])_{n\in\mathbb{N}}$ and $(T^{+n}u(x)-nc[0])_{n\in\mathbb{N}}$ are constantly equal to $u(x)$. This implies that $x \in \mathcal{A}$ (see Corollary 2.4.10) and $A_1 \subset \mathcal{A}$.

Assume $(x, y) \in A_2$. Then for any $u \in \mathcal{S}$, summing up the inequalities $u(y) - u(x) \leqslant c(x, y) + c[0]$ and $u(x) - u(y) \leqslant h(y, x)$ brings that $0 \leqslant c(x, y) + c[0] + h(y, x)$. As this is an equality, both inequalities were equalities and $(x, y) \in \widehat{\mathcal{A}}$. So $A_2 \subset \widehat{\mathcal{A}}$.

□

In order to obtain the reverse inclusions, we will use the following:

Proposition 3.1.6 *Let $x \in X$, then $T^{+n}h_x(x) - nc[0] \to 0$ as $n \to +\infty$.*

Proof Recall $T^{+n}h_x(x) - nc[0]$ is a non increasing sequence ($h_x \in \mathcal{S}$). Moreover, as $h_x(y) - T^{+n}h_x(x) + nc[0] \leqslant h(x, y)$, we see that $T^{+n}h_x(x) - nc[0] \geqslant 0$ for all $n > 0$ and the limit meets the same property.

For each $n > 0$, let $x = x_0^n, \cdots, x_n^n$ verify $T^{+n}h_x(x) = h_x(x_n^n) - \sum\limits_{i=0}^{n-1} c(x_i^n, x_{i+1}^n)$. Up to extracting, assume $x_{n_k}^{n_k} \to y$ for some y, passing to the limit, by definition of h we obtain that:

$$\lim_{k \to +\infty} T^{+n_k}h_x(x) - n_k c[0] \leqslant h_x(y) - h(x, y),$$

which proves the Proposition. □

End of the Proof of Theorem 3.1.5 Let $x \in \mathcal{A}$, then $T^{+n}h_x(x) - nc[0]$ is constant, hence it is 0, so $h(x, x) = 0$.

Let $(x, y) \in \widehat{\mathcal{A}}$, then $h_y(y) - h_y(x) = c(x, y) + c[0]$. But as $y \in \mathcal{A}$, $h(y, y) = 0$ hence $c(x, y) + c[0] + h(y, x) = 0$. □

Theorem 3.1.5 gives a more concrete characterization of the Aubry set. Pairs (x, y) in the 2-Aubry set are starting pairs of arbitrarily long loops of points with arbitrarily small cost (for the cost $c + c[0]$).

As a final byproduct of the preceding proofs we obtain:

Proposition 3.1.7 *If $(x, y) \in \widehat{\mathcal{A}}$ then $c(x, y) + c[0] = h(x, y)$.*

More generally, if $(x_n)_{n\in\mathbb{Z}} \in \widetilde{\mathcal{A}}$, then

$$\forall m < n, \quad \sum_{k=m}^{n-1} c(x_k, x_{k+1}) + (n-m)c[0] = h(x_m, x_n).$$

Proof As the function h_x is a critical subsolution we obtain that if $(x, y) \in \widehat{\mathcal{A}}$, $h_x(y) - h_x(x) = c(x, y) + c[0]$. As $h(x, x) = 0$ we get $h(x, y) = c(x, y) + c[0]$.

For the second equality, we sum the equalities $h_{x_m}(x_{k+1}) - h_{x_m}(x_k) = c(x_k, x_{k+1}) + c[0]$ to obtain

$$h(x_m, x_n) = h_{x_m}(x_n) - h_{x_m}(x_m) = \sum_{k=m}^{n-1} c(x_k, x_{k+1}) + (n-m)c[0].$$

□

3.2 Examples of Points in the Aubry Sets

We will now be more specific about the type of points that belong to the Aubry set. Most of them appear as limit points of minimizing chains. The following lemma is most useful:

Lemma 3.2.1 *Let $u : X \to \mathbb{R}$ be a weak KAM solution, then for all $x \in X$ there exists an infinite sequence $(x_{-n})_{n\geqslant 0}$ with $x_0 = x$ such that*

$$\forall n \in \mathbb{N}, \quad u(x) = u(x_{-n}) + \sum_{i=-n}^{-1} c(x_i, x_{i+1}) + nc[0]. \tag{3.2}$$

We call such sequences, calibrating sequences for u.

Proof It is an immediate consequence of successive applications of the fact that for all $x \in X$ there is $y \in X$ such that $u(x) = u(y) + c(y, x) + c[0]$. □

Proposition 3.2.2 *Let $u : X \to \mathbb{R}$ be a weak KAM solution and $(x_n)_{n\geqslant 0}$ a sequence given by Lemma 3.2.1, then $\alpha\big((x_{-n})_{n\geqslant 0}\big) \subset \mathcal{A}$. If we set $(\xi_{-n}) = \big((x_{-n-1}, x_{-n})\big)_{n\geqslant 0}$ the sequence of pairs of successive points of $(x_{-n})_{n\geqslant 0}$, then $\alpha\big((\xi_{-n})_{n\geqslant 0}\big) \subset \widehat{\mathcal{A}}$. Here α denotes the α-limit sets, that is limits of converging subsequences x_{-n_k} with $n_k \to +\infty$.*

Proof Let $y \in \alpha\big((x_{-n})_{n\geqslant 0}\big)$. There exists an extraction $n_k \to +\infty$ such that $x_{-n_k} \to y$. Up to taking a further extraction if necessary, we may assume that $n_{k+1} - n_k \to +\infty$ is an increasing sequence as well.

It follows from (3.2) that

$$\forall k \in \mathbb{N}, \quad u(x_{-n_k}) - u(x_{-n_{k+1}}) = \sum_{i=-n_{k+1}}^{-n_k-1} c(x_i, x_{i+1}) + (n_{k+1} - n_k)c[0]. \tag{3.3}$$

Passing to the limit, the continuity of the functions at stake imply that $u(y) - u(y) = 0 \geqslant h(y, y)$, hence $h(y, y) = 0$.

The same proof holds for the second part of the proposition. Late us take an extraction $n_k \to +\infty$ such that ξ_{-n_k} converges to (y, y') and such that $n_{k+1} - n_k \to +\infty$ is an increasing sequence as well. Passing to the limit in (3.3), we get that $0 \geqslant c(y, y') + c[0] + h(y', y)$. This means $(y, y') \in \widehat{\mathcal{A}}$. □

Remark 3.2.3 The same results (with same proof) hold for ω-limit sets of analogous sequences $(x_n)_{n\geqslant 0}$ for positive weak KAM solutions.

If u is a subsolution, the same results also hold for α and ω limit sets of elements of $\widetilde{\mathcal{A}}_u$.

We may now state a fundamental property of the Aubry set:

Theorem 3.2.4 *Let u and v be respectively a weak KAM solution and a subsolution such that $u_{|\mathcal{A}} \geqslant v_{|\mathcal{A}}$, then $u \geqslant v$.*

Let u and v be two weak KAM solutions such that $u_{|\mathcal{A}} = v_{|\mathcal{A}}$, then $u = v$. We say $\mathcal{A}$ is a uniqueness set for the critical equation.

Proof Let $x_0 \in X$ and let $(x_{-n})_{n \geqslant 0}$ be a calibrating sequence for u (see (3.2)). It then comes

$$\forall n \in \mathbb{N}, \quad u(x_0) - u(x_{-n}) = \sum_{i=-n}^{-1} c(x_i, x_{i+1}) + nc[0],$$

$$v(x_0) - v(x_{-n}) \leqslant \sum_{i=-n}^{-1} c(x_i, x_{i+1}) + nc[0].$$

Subtracting yields $u(x_0) - v(x_0) \geqslant u(x_{-n}) - v(x_{-n})$. As limiting points of the sequence $(x_{-n})_{n \geqslant 0}$ are in the projected Aubry set $\mathcal{A}$, taking a suitable converging subsequence and passing to the limit brings $u(x_0) - v(x_0) \geqslant 0$ since for u and v, the same inequality holds on $\mathcal{A}$. This proves the first result as x_0 is arbitrary.

In the second case, by symmetry, the opposite inequality holds, and the result follows as x_0 was taken arbitrarily.

□

Actually, a reciprocal statement can be proven. Being a subsolution on the Aubry set is the only obstruction to the existence of a weak KAM solution with prescribed values on $\mathcal{A}$.

Proposition 3.2.5 *Let $f : \mathcal{A} \to \mathbb{R}$ be a function such that $f(y) - f(x) \leqslant h(x, y)$ for all x and y in $\mathcal{A}$. Then there exists a weak KAM solution u such that $u_{|\mathcal{A}} = f$.*

Proof Let us check that the function u defined by $u(x) = \inf_{y \in \mathcal{A}} f(y) + h(y, x)$ satisfies the requirements. Let $x \in \mathcal{A}$. Then $f(y) + h(y, x) \geqslant f(x) - h(y, x) + h(y, x)$ by hypothesis. So $u(x) = f(x)$, by taking $y = x$ in the definition of u. The fact that u is a weak KAM solution is a consequence of the general fact that an infimum of weak KAM solutions is a weak KAM solution applied to the family of functions $f(y) + h_y$. This fact is proved in the next lemma. □

Lemma 3.2.6 *Let $(u_\alpha)_{\alpha \in A}$ be a family of subsolutions. Set $\overline{u} = \sup_{\alpha \in A} u_\alpha$ and $\underline{u} = \inf_{\alpha \in A} u_\alpha$. Then given that those functions are well defined, the following assertions hold:*

(i) The functions $\overline{u}$ and $\underline{u}$ are subsolutions.
(ii) If all the u_α are weak KAM solutions then $\underline{u}$ is a weak KAM solution.

Proof By definition,

$$\begin{aligned} T^{-}\underline{u}(x) &= \inf_{y\in X}\inf_{\alpha\in A} u_\alpha(y) + c(y,x) \\ &= \inf_{\alpha\in A}\inf_{y\in X} u_\alpha(y) + c(y,x) \\ &= \inf_{\alpha\in A} T^{-}u_\alpha(x) \geqslant \inf_{\alpha\in A} u_\alpha(x) - c[0] = \underline{u}(x) - c[0]. \end{aligned}$$

It was used that two infimums commute and that the u_α are subsolutions. Hence $\underline{u} \in \mathcal{S}$. Moreover if all the u_α are weak KAM solutions, then the inequality above turns out to be an equality and $\underline{u}$ is a weak KAM solution.

In a similar manner,

$$\begin{aligned} T^{-}\overline{u}(x) &= \inf_{y\in X}\sup_{\alpha\in A} u_\alpha(y) + c(y,x) \\ &\geqslant \sup_{\alpha\in A}\inf_{y\in X} u_\alpha(y) + c(y,x) \\ &= \sup_{\alpha\in A} T^{-}u_\alpha(x) \leqslant \sup_{\alpha\in A} u_\alpha(x) - c[0] = \overline{u}(x) - c[0]. \end{aligned}$$

□

3.3 Regularity of Subsolutions

In our discrete setting, the Aubry set enjoys the additional feature to be a set where all subsolutions present systematic regularity properties. Away from the Aubry set, this is false; subsolutions may fail to be continuous as was shown in [202]:

Theorem 3.3.1 *Let $x \in X$ be a non-isolated point. Then $x \in \mathcal{A}$ if and only if all subsolutions $u \in \mathcal{S}$ are continuous at x.*

Proof If $x \in \mathcal{A}$ then we have seen that $T^{+}u - c[0] \leqslant u \leqslant T^{-}u + c[0]$ and that equalities hold at x. As the left and right terms of the inequalities are continuous, then the famous Sandwich theorem implies that u is continuous at x.

The hypothesis that x is not isolated is used in the reciprocal (were x isolated, u could not be discontinuous at x). Assume now $x \notin \mathcal{A}$. Let u_0 be the continuous subsolution constructed in Theorem 2.4.1. In particular, the inequality $u \leqslant T^{-}u + c[0]$ is now strict at x and both functions are continuous. It follows from the in-between lemma (proved after) that any function v such that $u_0 \leqslant v \leqslant T^{-}u_0 + c[0]$ will be a subsolution. In particular, it can be constructed discontinuous at x, for example, take $u_0 = v$ everywhere, except at x and $v(x) = T^{-}u_0(x) + c[0]$. □

Lemma 3.3.2 *Let u be a subsolution and $v : X \to \mathbb{R}$ such that $u \leqslant v \leqslant T^{-}u + c[0]$, then $v \in \mathcal{S}$.*

Proof It follows from the monotonicity of the Lax–Oleinik semigroup:

$$u \leqslant v \leqslant T^- u + c[0] \leqslant T^- v + c[0].$$

□

3.4 More Regularity for Subsolutions

Of course, if more precise regularity results on subsolutions are aimed for, some structure has to be added. Until further notice, we will take as base space a compact smooth manifold M and c will be a cost on $M \times M$. Let us recall a fundamental definition first:

Definition 3.4.1

(i) Let $\Omega \subset \mathbb{R}^n$ be a convex open set. A function $f : \Omega \to \mathbb{R}$ is said to be K-*semiconcave* if the function $x \mapsto f(x) - K\|x\|^2$ is concave (the norm used is the Euclidean one). A function is *semiconcave*[1] if it is K-semiconcave for some $K \in \mathbb{R}$.

(ii) A function $f : \Omega \to \mathbb{R}$ is *locally semiconcave* if each $x \in \Omega$ belongs to a neighborhood V_x such that the restriction $f_{|V_x}$ is semiconcave.

(iii) A function $f : M \to \mathbb{R}$ is *locally semiconcave* if for all coordinate patch $\varphi : U \subset \mathbb{R}^n \to M$, the function $f \circ \varphi$ is locally semiconcave (n is assumed to be the dimension of M).

Remark 3.4.2 The property of being locally semiconcave is invariant by C^2 diffeomorphisms. Therefore, in the previous definition, it is enough that $f \circ \varphi_i$ be locally semiconcave for $\varphi_i : U_i \to M$ where the $\varphi_i(U_i)$ form any finite open cover of M. Of course, this property is much easier to establish.

As concave functions can be characterized as functions whose graphs admit a hyperplane tangent from above at every point, the following is implied:

Proposition 3.4.3 *A function* $f : \Omega \to \mathbb{R}$ *is* K*-semiconcave if and only if for all* $x \in \Omega$*, there exists a linear form* $p_x \in \mathbb{R}^{n*}$ *such that*

$$\forall y \in \Omega, \quad f(y) \leqslant f(x) + p_x(y - x) + K\|y - x\|^2. \tag{3.4}$$

The same holds true for locally semiconcave functions, but restricting to a neighborhood of x only. For locally semiconcave functions on a manifold M, $p_x \in T_x^* M$ becomes an element of the cotangent fiber at x and the inequality is true in a chart.

[1] The notion we refer to here is sometimes called semiconcave with linear modulus. For more details and proofs that are omitted here see [67]. Other good references are also the unavoidable [109] and the appendix of [110].

We call *superdifferential* of a function f, assumed to be semiconcave (resp. locally semiconcave), at x $\big($denoted $\partial^+ f(x)\big)$ the set of p_x such that (3.4) holds (resp. in a neighborhood of x or in a chart).

We state without proof:

Proposition 3.4.4 *Let $f : M \to \mathbb{R}$ be locally semiconcave and $x \in \mathbb{R}$. Then $\partial^+ f(x)$ is not empty, closed and convex. Moreover, f is differentiable at x if and only if $\partial^+ f(x)$ is a singleton (which then contains only $D_x f$).*

A very easy, though important, property of semiconcave functions is:

Proposition 3.4.5 *If $(f_\alpha)_{\alpha\in A}$ is a family of K-semiconcave functions on $U \subset \mathbb{R}^n$ then $\inf\limits_{\alpha\in A} f_\alpha$ is K-semiconcave as soon as it is well defined.*

The proof follows barely the analogous property of concave functions. Of course, each notion defined previously has a *semiconvex* counterpart which is defined by replacing concave by convex and $-$ signs by $+$. The opposite of a semiconcave function is then semiconvex and vice versa. A semiconvex function f has a *subdifferential* at each point denoted by $\partial^- f(x)$. It coincides with $-\partial^+(-f)(x)$.

Motivated by the above, one defines sub- and superdifferentials for general functions:

Definition 3.4.6 Let $f : U \to \mathbb{R}$ be a function defined on an open set of $\mathbb{R}^n$ the *superdifferential* $\partial^+ f(x)$ $\big($resp. *subdifferential* $\partial^- f(x)\big)$ of f at $x \in U$ is the set of $D_x\varphi$ where $\varphi : U \to \mathbb{R}$ is differentiable at x and verifies that $\varphi \geqslant f$ (resp. $\varphi \leqslant f$) with equality at x.

Remark 3.4.7 In the previous definition, the functions φ can be taken equivalently C^1. Sub- and superdifferentials are convex and closed. Moreover, f is differentiable at x if and only if they are both non-empty. In this case, $\partial^+ f(x) = \partial^- f(x) = \{D_x f\}$. As a locally semiconcave function has non-empty superdifferentials, we infer that if f is locally semiconcave, then f is differentiable at x if and only if $\partial^- f(x)$ is non-empty.

Here is a not so obvious property that explains the nature of some results:

Proposition 3.4.8 *A function $f : M \to \mathbb{R}$ is $C^{1,1}$ (differentiable with Lipschitz differential) if and only if it is both locally semiconcave and locally semiconvex.*

Definition 3.4.9 A family of functions $f_\alpha : M \to \mathbb{R}$ for $\alpha \in A$ is said equi-locally semiconcave if M can be covered by finitely many open charts $\varphi_i(U_i)$, where $U_i \subset \mathbb{R}^n$ and if there are constants K_i such that all $f_\alpha \circ \varphi_i$ are K_i-semiconcave.

Hypothesis
In the rest of this section, we assume that the families $c(x, \cdot)$ and $c(\cdot, x)$ for $x \in M$ are equi-locally semiconcave.

It can be checked (as M is compact) that a particular case of this is when the function c is itself locally semiconcave on $M \times M$.

When they exist, $\partial_1 c(x, y)$ and $\partial_2 c(x, y)$ denote the partial derivatives of c at (x, y) with respect to the first and second variable.

Coming back to discrete weak KAM theory, a refined version of Proposition 2.1.3 (i) then becomes:

Proposition 3.4.10 *Under the previous hypotheses, the image of T^- (resp. T^+) consists of equi-locally semiconcave (resp. equi-locally semiconvex) functions.*

The proof is nothing but a direct application of Proposition 3.4.5. Note that the result for T^+ involves semiconvexity because of the minus sign in its definition.

As consequences, let us derive some further regularizing properties of the Lax–Oleinik semigroups:

Proposition 3.4.11 *Let $v : M \to \mathbb{R}$ be a continuous function and $x_0 \in M$. Let $y_0 \in M$ verify that $T^-v(x_0) = v(y_0) + c(y_0, x_0)$ $\big($resp. $T^+v(x_0) = v(y_0) - c(x_0, y_0)\big)$. Then*

- *$\partial_2^+ c(y_0, x_0) \subset \partial^+ T^- v(x_0)$ (resp. $-\partial_1^+ c(x_0, y_0) \subset \partial^- T^+ v(x_0)$).*[2]
- *In particular, if T^-v (resp. T^+v) is differentiable at x_0 then $\partial_2 c(y_0, x_0)$ (resp. $-\partial_1^+ c(x_0, y_0)$) exists and $D_{x_0} T^- v = \partial_2 c(y_0, x_0)$ (resp. $-\partial_1^+ c(x_0, y_0) \subset D_{x_0} T^+ v$).*
- *If v is locally semiconcave (resp. semiconvex) then $D_{y_0} v = -\partial_1 c(y_0, x_0)$ (resp. $D_{y_0} v = \partial_2 c(x_0, y_0)$) and all the previous quantities do exist.*

Proof We prove half of the results leaving the rest as an exercise.

The first point is a direct consequence of the inequality

$$\forall x \in M, \quad T^- v(x) \leqslant v(y_0) + c(y_0, x),$$

which is an equality for $x = x_0$.

The second point then stems from the proved inclusion $\partial_2^+ c(y_0, x_0) \subset \partial^+ T^- v(x_0)$. By hypothesis, the right hand side is a singleton and the left hand side is not empty, hence they coincide and we get the result.

For the last part, note that the function $\varphi : y \mapsto v(y) + c(y, x_0)$ reaches its minimum at y_0. Hence $0 \in \partial^- \varphi(y_0)$. But by hypothesis, φ is locally semiconcave and it is easily verified that $\partial^+ v(y_0) + \partial_1^+ c(y_0, x_0) \subset \partial^+ \varphi(y_0)$. We infer that φ is differentiable at y_0 with $D_{y_0} \varphi = 0$ and that necessarily, $\partial^+ v(y_0)$ and $\partial_1^+ c(y_0, x_0)$ are singletons. Hence the result. □

Remark 3.4.12 The following results were actually proven and used: if $f : M \to \mathbb{R}$ and $g : M \to \mathbb{R}$ are locally semiconcave functions then

- $f + g$ is differentiable at some $x \in M$ if and only if both f and g are;

[2] By $\partial_2^+ c(y_0, x_0)$ we mean the superdifferential of the map $x \mapsto c(y_0, x)$ at x_0.

- if $f+g$ reaches a local minimum at some $x \in M$, then f and g are differentiable at x.

We have now the necessary material to state a more precise version of Theorem 3.3.1:

Theorem 3.4.13 *Let $x \in \mathcal{A}$. Then any subsolution u is differentiable at x. Moreover, $D_x u$ does not depend on u.*

Proof We will use the same inequalities as in Theorem 3.3.1. Let $u \in \mathcal{S}$, then $T^+u - c[0] \leqslant u \leqslant T^-u + c[0]$. Moreover those inequalities are equalities at $x \in \mathcal{A}$. This proves that both $\partial^- u(x) \neq \varnothing$ (as T^+u is locally semiconvex) and $\partial^+ u(x) \neq \varnothing$ (as T^-u is locally semiconcave). Hence u is differentiable at x. Note that T^-u and T^+u being subsolutions they are also differentiable at x and the first inequality above implies that all differentials are equal at x: $D_x u = D_x T^- u = D_x T^+ u$.

It remains to compute this differential. Let $(x', x) \in \widehat{\mathcal{A}}$. We know that $u(z) \leqslant u(x') + c(x', z) + c[0]$ for all $z \in M$ and equality holds at $z = x$. As u is differentiable at x, this implies that the function $u(x') + c(x', \cdot) + c[0]$ has a subdifferential at x, as it is locally semiconcave, it is differentiable and its differential is $\partial_2 c(x', x)$. Hence we conclude that $D_x u = \partial_2 c(x', x)$ which happens to be independent on $u \in \mathcal{S}$. □

Remark 3.4.14 A similar proof implies that if $(x, y) \in \widehat{\mathcal{A}}$ and $u \in \mathcal{S}$ then $D_x u = -\partial_1 c(x, y)$. But this is not surprising, as x verifies $c(x', x) + c(x, y) = \min_{z \in M} c(x', z) + c(z, y)$.

Let us stress one more time that c admits partial derivatives on the 2-Aubry set, as was actually established.

We now turn to improving Theorem 2.4.1:

Theorem 3.4.15 *There exists a strict subsolution u_1 which is $C^{1,1}$.*

The proof makes crucial use of Ilmanen's lemma (see [45, 54, 117, 132]).

Theorem 3.4.16 (Ilmanen's Lemma) *Given two functions $f, g : M \to \mathbb{R}$ such that f is locally semiconvex, g is locally semiconcave and $f \leqslant g$, there exists a function h which is $C^{1,1}$ such that $f \leqslant h \leqslant g$.*

Moreover, if h_0 is a continuous function such that $f \leqslant h_0 \leqslant g$, then h can be constructed arbitrarily close from h_0.

Proof of Theorem 3.4.15 The proof splits into two steps. First we construct $C^{1,1}$ subsolutions, then we explain how to make them strict.

Let us start with a subsolutions u. Then we have seen that $g = T^-u$ is a locally semiconcave subsolution, $f = T^+g - c[0]$ is a locally semiconvex subsolution and $f \leqslant g$. By Ilmanen's lemma, there exists a $C^{1,1}$ function h such that $f \leqslant h \leqslant g$. But the in between Lemma 3.3.2, transposed to T^+, tells us that $h \in \mathcal{S}$.

Now that we have a general procedure to construct subsolutions, let us see how to make them strict. Let u_0 be the continuous strict subsolution given by Theorem 2.4.1. Let us set $\varepsilon : M \times M \to \mathbb{R}$ the function defined by $\varepsilon(x, y) = c(x, y) + c[0] - u_0(y) + u_0(x)$. This function is everywhere non-negative and verifies $\varepsilon^{-1}\{0\} = \widehat{\mathcal{A}}$ thanks to the strictness property enjoyed by u_0. Let now ε_1 be a C^∞ function such that $0 \leqslant \varepsilon_1 \leqslant \varepsilon$ and $\varepsilon_1^{-1}\{0\} = \widehat{\mathcal{A}}$. Let us finally consider $\tilde{c} = c - \varepsilon_1$. This new cost still verifies that the marginal functions $\tilde{c}(x, \cdot)$ and $\tilde{c}(\cdot, y)$ are locally-uniformly semiconcave. Moreover, u_0 is a $c[0]$-subsolution[3] for $\tilde{c}$, indeed $u_0(y) - u_0(x) = c(x, y) + c[0] - \varepsilon \leqslant \tilde{c}(x, y) + c[0]$. The first part of the proof provides a $c[0]$-subsolution u_1 for this cost $\tilde{c}$ (using the semigroups $T^-_{\tilde{c}}$ and $T^+_{\tilde{c}}$ associated to $\tilde{c}$) which is $C^{1,1}$. Let us verify it is strict for c: for $(x, y) \in M \times M$, we compute

$$u_1(y) - u_1(x) \leqslant \tilde{c}(x, y) + c[0] = c(x, y) - \varepsilon_1(x, y) + c[0] \leqslant c(x, y) + c[0],$$

and this last inequality is strict whenever $(x, y) \notin \widehat{\mathcal{A}}$. This completes the proof. □

Remark 3.4.17 The previous Theorem can be made more precise. Using the fact that $T^+T^-u \leqslant u \leqslant T^-u + c[0]$ and the last assertion of Ilmanen's lemma, one proves that if u is continuous, then it can be approximated by $C^{1,1}$ strict subsolutions.

Finally, as a nontrivial convex combination of a subsolution with a strict subsolution is strict, one obtains that $C^{1,1}$ strict subsolutions are dense in the set $\mathcal{S} \cap C^0(M, \mathbb{R})$.

Let us also stress that, as pointed out in [54], Ilmanen's lemma (Theorem 3.4.16) can be recovered from Theorem 3.4.15 by considering the cost $c_{f,g}(x, y) = g(y) - f(x)$.

3.5 Graph Properties and Dynamics on the Aubry Set

Let us begin by mentioning a first general result under the hypotheses of the previous paragraph. A combination of Theorems 3.4.13 and 3.4.15 gives the following proposition which has a flavor of Mather's Graph Theorem:

Proposition 3.5.1 *There exists a set $\mathcal{A}^* \subset T^*M$ whose projection is $\mathcal{A}$ and such that if $(x, p) \in \mathcal{A}^*$ then any $u \in \mathcal{S}$ is differentiable at x and $D_x u = p$. Moreover the projection $\mathcal{A}^* \to \mathcal{A}$ is a bi-Lipschitz homeomorphism.*

Indeed, $\mathcal{A}^*$ is just the restriction of the graph of Du_1 to $\mathcal{A}$ where u_1 is given by Theorem 3.4.15.

In order to define a dynamics on the Aubry set, one would now like, given a point $x_0 \in \mathcal{A}$, to be able to reconstruct the whole sequence $(x_n)_{n\in\mathbb{Z}}$. To this aim, we impose an additional condition on the cost. It was studied in [201] and previously introduced in the setting of Optimal Transportation in [110]:

[3] Even though it can be proven that $c[0]$ is the critical constant for $\tilde{c}$, this fact is not useful in this proof.

Definition 3.5.2 (i) The cost c has the *left twist property* if for any y, the map $x \mapsto \partial_2 c(x, y)$ is injective on its domain of definition.[4]

(ii) The cost c has the *right twist property* if for any x, the map $y \mapsto \partial_1 c(x, y)$ is injective on its domain of definition.

(iii) The cost c enjoys the *twist condition* if it verifies both the left and the right twist properties.

(iv) We define the *left Legendre transform* $\mathcal{L}_\ell : \mathcal{D}_\ell \subset M \times M \to T^*M$ by $\mathcal{L}_\ell(x, y) = \big(y, \partial_2 c(x, y)\big)$ and the *right Legendre transform* $\mathcal{L}_r : \mathcal{D}_r \subset M \times M \to T^*M$ by $\mathcal{L}_r(x, y) = \big(x, -\partial_1 c(x, y)\big)$, where $\mathcal{D}_\ell$ and $\mathcal{D}_r$ are the sets of full measures on which the definitions make sense.

Under this twist condition, one gets this second version of Mather's Graph theorem:

Proposition 3.5.3 *Let us assume that c verifies the twist condition. Then both projections $\pi_i : \widehat{\mathcal{A}} \to \mathcal{A}$ are bijections.*

Proof We have seen in the proof of Theorem 3.4.13 and in the subsequent Remark that if $u \in \mathcal{S}$ and (x_{-1}, x_0, x_1) are successive points of a sequence $(x_n)_{n \in \mathbb{Z}}$ then $p = D_{x_0} u = \partial_2 c(x_{-1}, x_0) = -\partial_1 c(x_0, x_1)$. It follows from the left twist condition that $\mathcal{L}_\ell$ is injective and that $x_{-1} = \pi_1\big(\mathcal{L}_\ell^{-1}(x_0, p)\big)$ is uniquely determined. Similarly, it follows from the right twist condition that $\mathcal{L}_r$ is injective and that $x_1 = \pi_2\big(\mathcal{L}_r^{-1}(x_0, p)\big)$ is uniquely determined. □

Remark 3.5.4 In Optimal Transportation, similar twist conditions are used to prove existence of optimal transport maps for semiconcave costs. In the corresponding cases, such Optimal transport maps have their graphs included in analogues of the 2-Aubry set associated to Kantorovitch pairs. See the work of Fathi and Figalli for example [110].

3.6 Relations to the Classical Theory

This section comes back to the classical setting of a Tonelli Lagrangian L defined on the tangent bundle of a closed and compact smooth manifold M.

3.6.1 *The Classical Peierls Barrier*

The Peierls barrier for Lagrangian systems was introduced by Mather in [167], inspired by the works of Aubry and Le Daeron for twist maps [27]:

[4] The cost c being locally Lipschitz, this map is defined almost everywhere.

Definition 3.6.1 The Peierls barrier is defined by

$$\forall(x, y) \in M, \quad h_L(x, y) = \liminf_{t\to+\infty} h_t(x, y) + t\alpha(0),$$

where h_t is the minimal action functional previously introduced in (2.5).

It follows from Fathi's theorem on the convergence of the Lax–Oleinik semigroup [107], that in this autonomous setting, the liminf is actually a limit. Note that this is not necessarily the case in a discrete setting, or a time periodic setting, as shown in [111]. Actually Skander Charfi constructs in his PhD examples of time periodic Tonelli Hamiltonians and smooth initial conditions such that the associated solution is smooth for all times, recurrent but does not converge nor is periodic [75].

Proposition 3.6.2

(i) h_L *is well defined and continuous;*
(ii) for any subsolution $u \in \mathcal{S}'$ *and* $(x, y) \in M$, $u(y) - u(x) \leqslant h_L(x, y)$, *in particular for all* $x \in X$, $h_L(x, x) \geqslant 0$*;*
(iii) for all $(x, y, z) \in M^3$ *and real number* $t > 0$ *the following inequalities hold:*

$$h_L(x, y) \leqslant h_L(x, z) + h_t(z, y) + t\alpha(0);$$

$$h_L(x, y) \leqslant h_t(x, z) + t\alpha(0) + h_L(z, y);$$

$$h_L(x, y) \leqslant h_L(x, z) + h_L(z, y); \tag{3.5}$$

(iv) for all $x \in X$, *the function* $h_x = h_L(x, \cdot)$ *is a weak KAM solution and the function* $h^x = -h_L(\cdot, x)$ *is a positive weak KAM solution.*

The proof is the same as that of Proposition 3.1.2. Actually the links between the classical Peierls barrier and the discrete one is made even clearer by the next Proposition:

Proposition 3.6.3 *Let* h *be the Peierls barrier associated to the cost function* h_1*. Then* $h = h_L$*.*

Proof Once again, the proof heavily relies on the convergence of the Lax–Oleinik semigroup for autonomous Tonelli Lagrangians. Indeed, let $x \in M$, and define $v = h_1(x, \cdot)$. Then it follows from the definitions that if $t > 1$,

$$\forall y \in M, \quad h_t(x, y) = S(t-1)v(y).$$

Whence,

$$\forall y \in M, \quad h_L(x, y) = \liminf_{t\to+\infty} S(t-1)v(y) + t\alpha(0) = \lim_{t\to+\infty} S(t-1)v(y) + t\alpha(0),$$

while

$$\forall y \in M, \quad h(x, y) = \liminf_{n\to+\infty} S(n-1)v(y)+n\alpha(0) = \lim_{n\to+\infty} S(n-1)v(y)+t\alpha(0),$$

and $h = h_L$. □

Of course, the classical Peierls barrier allows to recover the Aubry sets, as in the discrete case:

Theorem 3.6.4 *The following equalities hold:*

$$\mathcal{A} = \{x \in M, \quad h_L(x, x) = 0\},$$
$$\mathcal{A}^* = \{(x, p) \in T^*M, \quad x \in \mathcal{A}, \quad p = D_x h_x\}.$$

Note that the first equality of the previous Theorem is actually the original definition of Mather in [167].

3.6.2 Examples of Points in the Aubry Set

We start by reviewing links between weak KAM solutions and the Aubry set. Those results can now be interpreted as consequences of the analogous results proven in the discrete setting but they were historically obtained first by Albert Fathi.

Proposition 3.6.5 *Let u be a weak KAM solution, then for all $x \in M$, there exists a C^2 curve $\gamma_x : (-\infty, 0] \to M$ such that $\gamma_x(0) = x$ and*

$$\forall t > 0, \quad u(x) = u\big(\gamma_x(-t)\big) + \int_{-t}^{0} L\big(\gamma_x(s), \dot\gamma_x(s)\big)\mathrm{d}s + t\alpha(0).$$

Such curves are called calibrating for u.

Of course, a similar statement is also valid for positive weak KAM solutions. Calibrating curves carry points of the Aubry set in their closure (α-limit set):

Proposition 3.6.6 *Let u be a weak KAM solution, $x \in M$ and $\gamma_x : (-\infty, 0] \to M$ be a calibrating curve given by the previous proposition. If $y \in \alpha(\gamma_x)$ then $y \in \mathcal{A}$. Moreover, if $(y, v) \in \alpha(\gamma_x, \dot\gamma_x)$, then $(y, v) \in \mathcal{A}'$.*

Of course a similar statement holds for positive weak KAM solutions and we let the reader infer it.

Conversely, knowing a subsolution or weak KAM solution on the Aubry set is rich in consequences:

Theorem 3.6.7 *Let u and v be respectively a weak KAM solution and a subsolution such that $u_{|\mathcal{A}} \geqslant v_{|\mathcal{A}}$. Then $u \geqslant v$.*

Let u and v be two weak KAM solutions such that $u_{|\mathcal{A}} = v_{|\mathcal{A}}$. Then $u = v$.

Note that thanks to the parallels made between discrete and classical theories, this Theorem is weaker than Theorem 3.2.4, as there are more discrete subsolutions than classical ones.

Finally, let us recall the converse to this Theorem:

Proposition 3.6.8 *Let $f : \mathcal{A} \to \mathbb{R}$ be a function such that $f(y) - f(x) \leqslant h(x, y)$ for all x and y in $\mathcal{A}$. Then there exists a weak KAM solution u such that $u_{|\mathcal{A}} = f$.*

3.6.3 Regularity and More Regularity of Subsolutions

We review here regularity properties of classical subsolutions and weak KAM solutions. Most results were obtained by Fathi and Siconolfi in two founding papers [114, 115]. The proofs are much more intricate than for the discrete theory. Note that on a non-empty compact connected smooth manifold (of positive dimension!) there is no isolated point.

Theorem 3.6.9 *Let $u : M \to \mathbb{R}$ be a critical (classical) subsolution. Then u is Lipschitz continuous on M.*

Let $x \in M$, then $x \in \mathcal{A}$ if and only if all critical subsolutions $u \in \mathcal{S}'$ are differentiable at x.

In Sect. 3.4 were introduced further assumptions on the underlying space and the cost function we work with. This originates in the following properties of the action functional and Lax–Oleinik semigroup in the classical theory:

Theorem 3.6.10 *For all $t > 0$, the minimal action functional h_t is semiconcave on $M \times M$.*

Let $u : M \to \mathbb{R}$ be a bounded function, then for all $t > 0$ the function $S^-(t)u$ is semiconcave and $S^+(t)u$ is semiconvex.

Given that the Lagrangian is C^2 and the functions h_t and Lax–Oleinik semigroups are defined by using infimum and supremum, the preceding Theorem may seem a posteriori natural (given results such as Proposition 3.4.5). Its consequences are very powerful.

For instance let us come back to Theorem 2.5.16. In the first version, Fathi and Siconolfi prove the existence of C^1 subsolutions by carefully studying the regularity of subsolutions on $\mathcal{A}$ and by a precise combination of smoothing and partitions of unity on $M \setminus \mathcal{A}$. Patrick Bernard instead has a more global and decisive idea establishing the following regularization result [42]:

Theorem 3.6.11 *Let $u : M \to \mathbb{R}$ be a bounded function and $t > 0$. There exists $\varepsilon > 0$ such that for all $t' < \varepsilon$, $S^-(t') \circ S^+(t)u$ is $C^{1,1}$.*

One should have in mind that a $C^{1,1}$ function is one that is both semiconcave and semiconvex (Proposition 3.4.8). So the idea behind the previous Theorem, and what

Bernard proves, is that the image of a semiconvex function by S^- stays semiconvex for small times. This is a general version of an older result known as Lasry–Lions regularization [149].

Then it should not come as a surprise that in our proof of Theorem 3.4.15, which is the discrete version of Bernard's Theorem, we used a composition of both operators T^- and T^+.

3.6.4 Graph Properties, Twist Condition and Dynamics on the Aubry Set

Let us start by noticing that thanks to the previous analysis, the Aubry set $\mathcal{A}^*$ introduced in the Lagrangian setting following Theorem 2.5.16 coincides with the set $\mathcal{A}^*$ introduced in Proposition 3.5.1 when applied to the cost function h_1. This explains the similar notation. Hence the conclusions of Proposition 3.5.1 also hold in our Lagrangian setting as they also follow from Bernard's Theorem 2.5.16. This is the content of Mather's Graph Theorem:

Proposition 3.6.12 *The projections $\mathcal{A}^* \to \mathcal{A}$ and $\mathcal{A}' \to \mathcal{A}$ are bi-Lipschitz homeomorphisms.*

This Chapter ends by explaining why the cost h_1 associated to a Tonelli Lagrangian satisfies the left and right twist conditions. This is presented in [201] and more details are given in [67, 109].

Proposition 3.6.13 *Let $L : TM \to \mathbb{R}$ be a Tonelli Lagrangian. Then the time-1 minimal action functional $h_1 : M \times M \to \mathbb{R}$ satisfies the left and right twist conditions.*

Proof Let $(x, y) \in M \times M$. Let $\gamma : [0, 1] \to M$ verify that

$$h_1(x, y) = \int_0^1 L\big(\gamma(s), \dot{\gamma}(s)\big)\mathrm{d}s,$$

with $\gamma(0) = x$ and $\gamma(1) = y$. Such a curve exists by Tonelli's Theorem. It is C^2 as already observed and solves the Euler–Lagrange equation. By standard variational arguments, one shows that $\big(-\partial_v L\big(x, \dot{\gamma}(0)\big), \partial_v L\big(y, \dot{\gamma}(1)\big)\big) \in \partial^+ h_1(x, y)$. It follows that if $\partial_1 h_1(x, y)$ exists, then $\partial_1 h_1(x, y) = -\partial_v L\big(x, \dot{\gamma}(0)\big)$. Remember that the Fenchel transform $\mathcal{L}$ defined by (2.2) is a diffeomorphism and observe that

$$\big(x, \partial_v L\big(x, \dot{\gamma}(0)\big)\big) = \mathcal{L}\big(x, \dot{\gamma}(0)\big) = \big(x, -\partial_1 h_1(x, y)\big).$$

We deduce that the preceding equation uniquely determines $\dot\gamma(0)$ and that the minimizing curve γ is unique with $\big(\gamma(s), \dot\gamma(s)\big) = \varphi_L^s\big(\gamma(0), \dot\gamma(0)\big)$ for $s \in [0, 1]$. Another consequence is that, denoting by $\pi : TM \to M$ the canonical projection,

$$\pi \circ \varphi_L^1 \circ \mathcal{L}^{-1}\big(x, -\partial_1 h_1(x, y)\big) = \pi \circ \varphi_L^1\big(x, \dot\gamma(0)\big) = y.$$

Hence $y \mapsto -\partial_1 h_1(x, y)$ is injective, and h_1 has the right twist property. The proof of the left twist property is exactly the same.

□

We may also interpret the left and right Legendre transforms introduced in Definition 3.5.2. Indeed, if $(x, y) \in \mathcal{D}_r$ then if $(x, v) = \mathcal{L}^{-1} \circ \mathcal{L}_r(x, y)$, v is the initial speed of the unique minimizing curve, going from x to y in time 1. This has the following consequence relating different Aubry sets:

Proposition 3.6.14 *The following equalities hold for the cost h_1:*

$$\widehat{\mathcal{A}} = \Big\{\Big(x, \pi \circ \varphi_L^1(x, v)\Big), \quad (x, v) \in \mathcal{A}'\Big\},$$
$$\widetilde{\mathcal{A}} = \Big\{\big(\pi \circ \varphi_L^n(x, v)\big)_{n\in\mathbb{Z}}, \quad (x, v) \in \mathcal{A}'\Big\}.$$

Moreover, for all $(x, v) \in \mathcal{A}'$, $h_1\big(x, \pi \circ \varphi_L^1(x, v)\big) = \int_0^1 L \circ \varphi_L^s(x, v)\mathrm{d}s$.

Finally, let us mention a curious fact about h_1. It can be established, using the notions of reachable gradient, that $\partial_1 h_1(x, y)$ exists if and only if there exists a unique minimizing curve, in time 1, going from x to y (see [201]). Similarly, $\partial_2 h_1(x, y)$ exists if and only if there exists a unique minimizing curve, in time 1, going from x to y. It is therefore obtained that for the particular cost h_1, the equality $\mathcal{D}_\ell = \mathcal{D}_r$ holds.

Chapter 4
Minimizing Mather Measures and the Discounted Semigroups

Abstract This central chapter revolves around Mather measures. We give two equivalent definitions of those measures, prove their existence and relate their support with the Aubry sets. We then use those measures to prove the convergence of solutions to the discounted equations. This convergence result is also generalized to degenerate discounted equations. Finally we study similarities and differences of discrete and continuous Mather measures in the classical Lagrangian setting. The same is done for the discrete and the continuous discounted equations.

In this part we go back to the more general setting of a continuous cost c on a compact metric space X. Most results were presented and used by the authors and Fathi, Iturriaga, Davini in [94] to study convergence of solutions of the discounted equations. Earlier results and the introductions of Mather measures had appeared in Bernard and Buffoni's work [49].

The study of the positive counterpart to the discounted equations is new to our knowledge, both in the discrete and in the continuous setting. So are the results concerning degenerate discounted equations in the discrete setting.[1]

4.1 Minimizing Mather Measures

The cost c is a continuous function from $X \times X$ to $\mathbb{R}$ and both canonical projections from $X \times X$ to X are denoted π_1 and π_2.

[1] Between the time the first version of this text was written and its publication, the degenerate versions were widely generalized in [180].

M. Zavidovique, *Discrete Weak KAM Theory*, Lecture Notes in Mathematics 2377,
https://doi.org/10.1007/978-3-031-96809-9_4

4.1.1 An Optimal Transport Like Approach

Recall that if μ is a Borel measure on $X \times X$ then $\pi_{1*}\mu$ and $\pi_{2*}\mu$ are measures on X defined as follows: if $A \subset X$ is a Borel set, then $\pi_{1*}\mu(A) = \mu(A \times X)$ and $\pi_{2*}\mu(A) = \mu(X \times A)$.

Definition 4.1.1 A Borel probability measure μ on $X \times X$ is said to be *closed* if it has equal marginals: $\pi_{1*}\mu = \pi_{2*}\mu$. We will denote by $\widehat{\mathcal{P}}$ the set of closed Borel probability measures on $X \times X$.

The previous condition is equivalent to the following:

Proposition 4.1.2 *A probability measure μ is closed if and only if for any continuous function $f : X \to \mathbb{R}$, $\int_{X\times X} \big(f(y) - f(x)\big)\mathrm{d}\mu(x, y) = 0$.*

Proof The proof is left as an exercise but follows these lines: if μ is closed, then the property of the proposition holds for indicatrix functions of open or closed sets. Hence it holds for simple functions (linear combination of indicatrix functions) and by density, it holds for continuous functions.

The converse is proved by approximating (from above and below) indicator functions by continuous functions. □

The set $\widehat{\mathcal{P}}$ of closed probability measures is clearly convex, closed and compact (for the weak $*$ topology).

Examples of closed measures can be constructed using Birkhoff averages. Indeed, given $(x_1, \cdots, x_n) \in X^n$, the measure $\mu = \frac{1}{n}\sum\limits_{i=1}^{n} \delta_{(x_i,x_{i+1})}$,[2] with the convention that $x_{n+1} = x_1$, is closed. Its marginals are

$$\pi_{1*}\mu = \frac{1}{n}\sum_{i=1}^{n} \delta_{x_i} = \frac{1}{n}\sum_{i=1}^{n} \delta_{x_{i+1}} = \pi_{2*}\mu.$$

Let us now introduce the concept of *minimizing measure*. It was first introduced by Mather for twist maps in [163] and studied by Bernard and Buffoni in [49] in a context similar to the present one, following their earlier works on optimal transportation [47, 48].

Theorem 4.1.3 *The following equality holds:*

$$-\, c[0] = \min_{\mu\in\widehat{\mathcal{P}}} \int_{X\times X} c(x, y)\mathrm{d}\mu(x, y).$$

[2] The notation δ stands for a Dirac mass.

There exists a closed measure realizing the minimum in the previous equality. Moreover, a closed measure realizes this minimum if and only if it is supported on the 2–Aubry set $\widehat{\mathcal{A}}$.

Proof Let $\mu \in \widehat{\mathcal{P}}$ and u_0 a strict continuous subsolution given by Theorem 2.4.1. Then one has

$$0 = \int_{X\times X} \big(u_0(y) - u_0(x)\big)\mathrm{d}\mu(x, y) \leqslant \int_{X\times X} \big(c(x, y) + c[0]\big)\mathrm{d}\mu(x, y).$$

This proves that $-c[0] \leqslant \min\limits_{\mu\in\widehat{\mathcal{P}}} \int_{X\times X} c(x, y)\mathrm{d}\mu(x, y)$. Moreover, as u_0 is continuous, one has equality for a measure μ if and only if $u_0(y) - u_0(x) = c(x, y) + c[0]$ for μ–almost–every (x, y) that is if μ is supported on $\widehat{\mathcal{A}}$.

Let us now construct such a measure. We use Birkhoff averages. Let $f : X \to \mathbb{R}$ be any continuous function, $x \in X$ and for all $n \in \mathbb{N}$, let $x^n_{-n}, \cdots, x^n_0 = x$ verify that $T^{-n} f(x) = f(x^n_{-n}) + \sum\limits_{i=-n}^{-1} c(x^n_i, x^n_{i+1})$. Define $\mu_n = \frac{1}{n} \sum\limits_{i=-n}^{-1} \delta_{(x^n_i, x^n_{i+1})}$. Finally let $(n_k)_{k\in\mathbb{N}}$ be an extraction[3] such that the $(\mu_{n_k})_{k\in\mathbb{N}}$ converge to a measure μ. As the μ_n are probability measures, so is μ.

Let us verify that μ is closed: this follows by passing to the limit in the inequality

$$\begin{aligned}\forall g \in C^0(X, \mathbb{R}),\quad & \Big|\int_{X\times X} \big(g(y) - g(x)\big)\mathrm{d}\mu_n(x, y)\Big| \\ &= \frac{1}{n}\Big|\sum_{i=-n}^{-1} g(x^n_{i+1}) - g(x^n_i)\Big| \\ &= \frac{1}{n}|g(x^n_{-n}) - g(x)| \leqslant \frac{2}{n}\|g\|_\infty \to 0.\end{aligned}$$

Let us verify μ is minimizing: recall that the family $T^{-n} f + nc[0]$ is uniformly bounded. Hence

$$\int_{X\times X} \big(c(x, y) + c[0]\big)\mathrm{d}\mu_n(x, y) = \frac{1}{n}\big(T^{-n} f(x^n_{-n}) + nc[0] - f(x)\big) \to 0.$$

This proves that $\int_{X\times X} c(x, y)\mathrm{d}\mu = -c[0]$ and concludes the proof. □

Definition 4.1.4 We denote by $\widehat{\mathcal{P}}_0$ the set of minimizing closed probability measures, that is, the set of closed probability measures $\mu \in \widehat{\mathcal{P}}$ such that $\int c(x, y)\mathrm{d}\mu = -c[0]$. Such a measure μ is termed a *Mather measure*.

[3] The set of Borel probability measures on a compact space is compact for the weak-$*$ topology.

We define the Mather set $\widehat{\mathcal{M}} \subset X \times X$ by

$$\widehat{\mathcal{M}} = \overline{\bigcup_{\mu \in \widehat{\mathcal{P}}_0} \mathrm{supp}(\mu)},$$

where supp stands for the support of a measure. The projected Mather set is

$$\mathcal{M} = \pi_1(\widehat{\mathcal{M}}) = \pi_2(\widehat{\mathcal{M}}).$$

Remark 4.1.5 The set $\widehat{\mathcal{P}}_0$ is clearly itself compact and convex. Moreover by Theorem 4.1.3 the Mather set is a subset of the 2–Aubry set: $\widehat{\mathcal{M}} \subset \widehat{\mathcal{A}}$.

Finally, the Mather set is by definition closed, but one can prove that there is no need to take the closure in its definition. Indeed, there exists one minimizing measure μ_0 whose support is the whole of $\widehat{\mathcal{M}}$. To construct it, one considers a sequence $(\mu_n)_{n>0}$ dense in $\widehat{\mathcal{P}}_0$ and one then verifies that $\mu_0 = \sum\limits_{n>0} \frac{1}{2^n}\mu_n$ meets all the requirements.

The proof of Theorem 4.1.3 sheds, once more, light on the general principle that long minimizing chains cannot stay too far from the Aubry set (as already seen in Proposition 3.2.2). This allows to give a stronger version of Theorem 3.2.4 and Proposition 3.2.5:

Theorem 4.1.6

1. *Let u and v be respectively a weak KAM solution and a subsolution such that $u_{|\mathcal{M}} \geqslant v_{|\mathcal{M}}$. Then $u \geqslant v$.*

   Let u and v be two weak KAM solutions such that $u_{|\mathcal{M}} = v_{|\mathcal{M}}$. Then $u = v$.
2. *Conversely, let $f : \mathcal{M} \to \mathbb{R}$ be a function such that $f(y) - f(x) \leqslant h(x, y)$ for all x and y in $\mathcal{M}$, where h is Peierl's barrier. Then there exists a weak KAM solution u such that $u_{|\mathcal{M}} = f$.*

Proof

1. Let $x_0 \in X$ and let $(x_{-n})_{n\geqslant 0}$ be a calibrating sequence for u. As observed in the proof of Theorem 3.2.4, $u(x_0) - v(x_0) \geqslant u(x_{-n}) - v(x_{-n})$ for all $n > 0$.

   Limiting points of the sequence $(x_{-n})_{n\geqslant 0}$ are not necessarily in $\mathcal{M}$. However, we prove that there exists a suitable subsequence converging to a point in $\mathcal{M}$, allowing to conclude the proof as in Theorem 3.2.4. Assume by contradiction the contrary. There exists an $\varepsilon > 0$ such that $d(x_{-n}, \mathcal{M}) \geqslant \varepsilon$ for all $n \in \mathbb{N}$. Let $F = \{x \in X,\ \ d(x, \mathcal{M}) \geqslant \varepsilon\}$, that is a closed set. As in the proof of Proposition 4.1.3, define $\mu_n = \frac{1}{n}\sum\limits_{i=-n}^{-1} \delta_{(x_i, x_{i+1})}$. Let finally $(n_k)_{k\in\mathbb{N}}$ be an extraction such that the sequence $(\mu_{n_k})_{k\in\mathbb{N}}$ converges to a measure μ. By hypothesis all the μ_n have their support included in $F \times F$, so the same holds for

μ. But, as proved in Proposition 4.1.3, $\mu \in \widehat{\mathcal{P}}_0$ is a Mather measure, hence the support of μ is included in $\widehat{\mathcal{M}}$, and this is absurd.

In the second case, by symmetry, the opposite inequality holds, and the result follows as x_0 was taken arbitrary.

2. This point is established exactly as 3.2.5, we do not reproduce its proof.

□

4.1.2 An Ergodic Point of View

Mather sets were introduced as subsets of $X \times X$. As for Aubry sets, that hides the underlying dynamics. As Aubry sets may be equivalently defined on $X \times X$ or on $X^{\mathbb{Z}}$, there are analogous measures defined on $X^{\mathbb{Z}}$. This is explained in [49, Paragraph 4.2]. Indeed, denoting $s : (x_n)_{n\in\mathbb{Z}} \mapsto (x_{n+1})_{n\in\mathbb{Z}}$ the shift operator, given a Borel probability measure $\tilde{\mu}$ on $X^{\mathbb{Z}}$ that is invariant by s, its push-forward $(\pi_{0,1})_*\tilde{\mu}$ by the projection $\pi_{0,1} : (x_n)_{n\in\mathbb{Z}} \mapsto (x_0, x_1)$ is a Borel probability measure on $X \times X$ that is closed in the sense of Definition 4.1.1 and such that $\int_{X^{\mathbb{Z}}} c(x_0, x_1)\mathrm{d}\tilde{\mu}\big((x_n)_{n\in\mathbb{Z}}\big) = \int_{X\times X} c(x, y)\mathrm{d}(\pi_{0,1})_*\tilde{\mu}$.

Conversely, if μ is a Borel closed probability measure on $X \times X$, Bernard and Buffoni construct, via a disintegration of μ with respect to the projection $\pi_2 : X \times X \to X$, a shift invariant measure $\tilde{\mu}$ on $X^{\mathbb{Z}}$ such that $\int_{X^{\mathbb{Z}}} c(x_0, x_1)\mathrm{d}\tilde{\mu}\big((x_n)_{n\in\mathbb{Z}}\big) = \int_{X\times X} c(x, y)\mathrm{d}\mu$. We therefore derive the following analogues of previous results, either by using the correspondence of Bernard and Buffoni, or by reproducing the proofs in this context. We leave the latter to the reader.

Definition 4.1.7 Denote $\widetilde{\mathcal{P}}$ be the set of shift invariant Borel probability measures on $X^{\mathbb{Z}}$. This is the set of Borel probability measures $\tilde{\mu}$ on $X^{\mathbb{Z}}$ such that $s_*\tilde{\mu} = \tilde{\mu}$.

The following result holds:

Proposition 4.1.8 *The critical constant is characterized by*

$$\inf_{\tilde{\mu}\in\widetilde{\mathcal{P}}} \int_{X^{\mathbb{Z}}} c(x_0, x_1)\mathrm{d}\tilde{\mu}\big((x_n)_{n\in\mathbb{Z}}\big) = \min_{\tilde{\mu}\in\widetilde{\mathcal{P}}} \int_{X^{\mathbb{Z}}} c(x_0, x_1)\mathrm{d}\tilde{\mu}\big((x_n)_{n\in\mathbb{Z}}\big) = -c[0].$$

Moreover, an invariant measure $\tilde{\mu} \in \widetilde{\mathcal{P}}$ is minimizing if and only if it is supported on the Aubry set $\widetilde{\mathcal{A}}$.

Definition 4.1.9 We define $\widetilde{\mathcal{P}}_0 \subset \widetilde{\mathcal{P}}$ to be the set of shift invariant Borel probability measures $\tilde{\mu}_0$ on $X^{\mathbb{Z}}$ such that

$$\int_{X^{\mathbb{Z}}} c(x_0, x_1)\mathrm{d}\tilde{\mu}_0\big((x_n)_{n\in\mathbb{Z}}\big) = \inf_{\tilde{\mu}\in\widetilde{\mathcal{P}}} \int_{X^{\mathbb{Z}}} c(x_0, x_1)\mathrm{d}\tilde{\mu}\big((x_n)_{n\in\mathbb{Z}}\big) = -c[0].$$

Such measures are also called *minimizing* or *Mather measures* and the context makes it clear whether a measure is defined on $X^{\mathbb{Z}}$ or on $X \times X$.

We define the Mather set $\widetilde{\mathcal{M}} \subset X^{\mathbb{Z}}$ by

$$\widetilde{\mathcal{M}} = \overline{\bigcup_{\tilde{\mu} \in \widetilde{\mathcal{P}}_0} \operatorname{supp}(\tilde{\mu})},$$

where supp stands for the support of a measure.

The set $\widetilde{\mathcal{P}}_0$ is convex and compact. Finally the initial discussion together with Proposition 2.4.5 yield that:

Proposition 4.1.10 *The following equalities hold:* $\mathcal{M} = \pi_0(\widetilde{\mathcal{M}})$ *and* $\widehat{\mathcal{M}} = \pi_{0,1}(\widetilde{\mathcal{M}})$.

4.2 The Discounted Equation

This Chapter ends by returning to the roots, more precisely to the second proof of the weak KAM Theorem 2.2.1. Recall that if $\lambda \in (0, 1)$ then u_λ is the unique function such that $u_\lambda = T_\lambda^- u_\lambda = T^-(\lambda u_\lambda)$. We now prove a result first obtained in [94]:

Theorem 4.2.1 *There exists a weak KAM solution* u_1 *such that* $u_\lambda + \frac{c[0]}{1-\lambda} \to u_1$ *where the convergence takes place as* $\lambda \to 1$ *and is uniform.*

The proof is divided into several steps. It was already shown that as $\lambda \to 1$, $(1-\lambda)u_\lambda \to -c[0]$ $\big($Remark 2.2.4 (ii)$\big)$. Actually one gets something more precise:

Proposition 4.2.2 *The family* $u_\lambda + \frac{c[0]}{1-\lambda}$ *is uniformly bounded as* $\lambda \to 1$.

This will be a simple consequence of the following comparison principle:

Lemma 4.2.3 *Let* v_1 *be such that* $v_1 \leqslant T_\lambda^- v_1$ *and let* v_2 *verify* $v_2 \geqslant T_\lambda^- v_2$. *Then* $v_1 \leqslant u_\lambda \leqslant v_2$.

Proof By induction, one has for all $n \in \mathbb{N}$ that $v_1 \leqslant T_\lambda^{-n} v_1$ and $v_2 \geqslant T_\lambda^{-n} v_2$. Both right hand side terms converge to u_λ as $n \to +\infty$ (recall T_λ^- is a contraction). The results follow by passing to the limit. □

Proof of Proposition 4.2.2 Let u be a weak KAM solution. Then adding and subtracting big constants to u provides two weak KAM solutions $\overline{u}$ and $\underline{u}$ which are positive and negative respectively hence verify $\overline{u} \geqslant \lambda\overline{u}$ and $\underline{u} \leqslant \lambda\underline{u}$. We then obtain that

$$\forall \lambda \in (0, 1), \quad \overline{u} - c[0] = T^-(\overline{u}) \geqslant T^-(\lambda\overline{u}) = T_\lambda^-(\overline{u}).$$

This can be rewritten $\overline{u} - \frac{c[0]}{1-\lambda} \geqslant T_\lambda^-(\overline{u} - \frac{c[0]}{1-\lambda})$. In a same manner, $\underline{u} - \frac{c[0]}{1-\lambda} \leqslant T_\lambda^-(\underline{u} - \frac{c[0]}{1-\lambda})$. Apply the previous lemma to obtain that $\underline{u} - \frac{c[0]}{1-\lambda} \leqslant u_\lambda \leqslant \overline{u} - \frac{c[0]}{1-\lambda}$ which implies the proposition. □

As the functions $u_\lambda + \frac{c[0]}{1-\lambda}$ are equicontinuous and equibounded, thanks to the Arzelà–Ascoli Theorem, to prove the convergence it is enough to prove that all converging subsequences have the same limit. We now establish constraints on such limits:

Proposition 4.2.4 *Let $\mu \in \widehat{\mathcal{P}}_0$ be a Mather measure. Assume $u_{\lambda_n} + \frac{c[0]}{1-\lambda_n} \to u$ as $n \to +\infty$ for some extraction $\lambda_n \to 1$. Then $\int_X u(x)\mathrm{d}\pi_{1*}\mu(x) \leqslant 0$.*

Proof Start from the inequalities $u_\lambda(y) - \lambda u_\lambda(x) \leqslant c(x, y)$ for all pairs (x, y). Integrating with respect to μ yields

$$\int_{X\times X} \big(u_\lambda(y) - \lambda u_\lambda(x)\big)\mathrm{d}\mu(x, y) \leqslant \int_{X\times X} c(x, y)\mathrm{d}\mu(x, y) = -c[0],$$

as μ is minimizing. But since μ is closed, both marginals are equals and the left hand side is equal to $(1-\lambda)\int_X u_\lambda(x)\mathrm{d}\pi_{1*}\mu(x)$. Dividing by $(1-\lambda)$ one obtains that $\int_X \big(u_\lambda(x) + \frac{c[0]}{1-\lambda}\big)\mathrm{d}\pi_{1*}\mu(x) \leqslant 0$. The result now follows taking $\lambda = \lambda_n$ and passing to the limit. □

Note that as the functions u_λ are equicontinuous, any accumulation point u as in the previous Proposition is automatically continuous.

The next step is to identify a reasonable candidate for the limit. This is done in the next Definition:

Definition 4.2.5 Let $\mathcal{F} \subset \mathcal{S} \cap C^0(X, \mathbb{R})$ be the set of continuous subsolutions u verifying the constraint $\int_X u(x)\mathrm{d}\pi_{1*}\mu(x) \leqslant 0$ for all Mather measures $\mu \in \widehat{\mathcal{P}}_0$.

We define $u_1 = \sup\limits_{u\in\mathcal{F}} u$ where the supremum is taken pointwise.

The set $\mathcal{F}$ is not empty for it contains negative subsolutions (recall $\mathcal{S}$ or the set of weak KAM solutions are invariant by addition of constants). Restricting to continuous functions is not necessary (see [94] for the alternative approach of considering all subsolutions), but it simplifies some proofs. Elements of $\mathcal{F}$ are bounded above as they must take at least a non–positive value. Hence u_1 is well defined. The idea of taking supremums of solutions or subsolutions in viscosity solutions theory is rather standard, we will see here that it is very useful.

Of course, Proposition 4.2.4 has a trivial consequence: if $u = \lim\limits_{n\to+\infty} u_{\lambda_n} + \frac{c[0]}{1-\lambda_n}$ for some sequence $\lambda_n \to 1$ then $u \in \mathcal{F}$ and $u \leqslant u_1$.

In order to establish the full convergence, we have to prove the reverse inequality. This will be done by constructing some appropriate Mather measures. First we give a representation formula for u_λ:

Lemma 4.2.6 *For any $\lambda \in (0, 1)$ and $x \in X$, we have*

$$u_\lambda(x) = \min_{\substack{(x_n)_{n\leqslant 0}\\ x_0=x}} \sum_{n\leqslant 0} \lambda^{-n} c(x_{n-1}, x_n).$$

Proof As T_λ^- is a contraction on the set of continuous functions its fixed point is the limit of the iterates starting with any initial function. Taking the 0 function, one computes that if $k > 0$,

$$T_\lambda^{-k}\bar{0}(x) = \min_{x_{-k},\cdots,x_0=x} \sum_{i=-k+1}^{0} \lambda^{-i} c(x_{i-1}, x_i).$$

The result follows letting $n \to +\infty$. The fact that all infimums are minimums comes from the usual compactness arguments. □

Proposition 4.2.7 *Let $x \in X$ and for all $\lambda \in (0, 1)$ let $(x_n^\lambda)_{n\leqslant 0}$ such that $x_0^\lambda = x$ and $u_\lambda(x) = \sum\limits_{n\leqslant 0} \lambda^{-n} c(x_{n-1}^\lambda, x_n^\lambda)$. Define the probability measure μ_λ by*

$$\forall f \in C^0(X \times X, \mathbb{R}), \quad \int_{X\times X} f(x, y)\mathrm{d}\mu_\lambda(x, y) = (1-\lambda)\sum_{n\leqslant 0} \lambda^{-n} f(x_{n-1}^\lambda, x_n^\lambda).$$

Assume finally that for some subsequence $\lambda_n \to 1$ the sequence $(\mu_{\lambda_n})_{n\in\mathbb{N}}$ converges to μ. Then the measure μ is a Mather measure.

Proof The multiplicative term $(1-\lambda)$ ensures that the measures μ_λ are probability measures. Hence so is μ. We therefore have to prove that μ is closed and minimizing.

The fact that μ is closed does not depend on the particular choice of the sequences $(x_n^\lambda)_{n\leqslant 0}$ and results from the following computation:

Let $f : X \to \mathbb{R}$ be a continuous function. Then

$$\begin{aligned}\Big|\int_{X\times X} \big(f(y) - f(x)\big)\mathrm{d}\mu_\lambda(x, y)\Big| &= (1-\lambda)\Big|\sum_{n\leqslant 0} \lambda^{-n}\big(f(x_n^\lambda) - f(x_{n-1}^\lambda)\big)\Big| \\ &= (1-\lambda)\Big|f(x) + \sum_{n\leqslant -1} (\lambda^{-n} - \lambda^{-n-1}) f(x_n^\lambda)\Big| \\ &\leqslant (1-\lambda)\|f\|_\infty\Big(1 + (1-\lambda)\sum_{n\leqslant -1} \lambda^{-n-1}\Big) \\ &= 2(1-\lambda)\|f\|_\infty \to 0.\end{aligned}$$

On the contrary, the fact that μ is minimizing depends heavily on the use of the definition of (x_n^λ):

$$(1-\lambda)u_\lambda(x) = (1-\lambda)\sum_{n\leqslant 0}\lambda^{-n}c(x_{n-1}^\lambda, x_n^\lambda) = \int_{X\times X} c(x,y)\mathrm{d}\mu_\lambda(x,y).$$

As $\lambda_n \to 1$, the left hand side goes to $-c[0]$ by Remark 2.2.4 (ii), and the right hand side converges to $\int_{X\times X} c(x,y)\mathrm{d}\mu(x,y)$. □

We now explain why those measures play a particular role:

Lemma 4.2.8 *Let $w \in \mathcal{S}$ be a continuous subsolution, then using the previous notation,*

$$\forall\lambda \in (0,1),\quad u_\lambda(x) + \frac{c[0]}{1-\lambda} \geqslant w(x) - \int_X w(z)\mathrm{d}\pi_{1*}\mu_\lambda(z).$$

Proof We start with the definition of u_λ and then use that w is a subsolution as follows:

$$\begin{aligned} u_\lambda(x) + \frac{c[0]}{1-\lambda} &= \sum_{n\leqslant 0}\lambda^{-n}\big(c(x_{n-1}^\lambda, x_n^\lambda) + c[0]\big) \geqslant \sum_{n\leqslant 0}\lambda^{-n}\big(w(x_n^\lambda) - w(x_{n-1}^\lambda)\big) \\ &= w(x) - \sum_{n\leqslant 0}(\lambda^{-n} - \lambda^{-n+1})w(x_{n-1}^\lambda) \\ &= w(x) - (1-\lambda)\sum_{n\leqslant 0}\lambda^{-n}w(x_{n-1}^\lambda) \\ &= w(x) - \int_X w(z)\mathrm{d}\pi_{1*}\mu_\lambda(z). \end{aligned}$$

□

At last, let us conclude:

Proof of Theorem 4.2.1 Let $u_{\lambda_n} + \frac{c[0]}{1-\lambda_n} \to u$ be a converging subsequence, we have already seen that $u \leqslant u_1$ where u_1 is given by Definition 4.2.5.

Let now $x \in X$ and for $\lambda \in (0,1)$, let $(x_n^\lambda)_{n\leqslant 0}$ such that $x_0^\lambda = x$ and $u_\lambda(x) = \sum_{n\leqslant 0}\lambda^{-n}c(x_{n-1}^\lambda, x_n^\lambda)$ and define the probability measure μ_λ as in Proposition 4.2.7. Extracting a further subsequence, assume that the μ_{λ_n} converge to a measure μ which is then a Mather measure by Proposition 4.2.7. Let $w \in \mathcal{F}$, applying the previous Lemma 4.2.8 we get $u_\lambda(x) + \frac{c[0]}{1-\lambda} \geqslant w(x) - \int_X w(z)\mathrm{d}\pi_{1*}\mu_\lambda(z)$ and along the subsequence λ_n letting $n \to +\infty$ yields (using $w \in \mathcal{F}$)

$$u(x) \geqslant w(x) - \int_X w(z)\mathrm{d}\pi_{1*}\mu(z) \geqslant w(x).$$

Taking the supremum over $w \in \mathcal{F}$, we conclude that $u(x) \geqslant u_1(x)$. Hence we have established the convergence.

□

As a byproduct of the previous proof and of Proposition 4.2.4, we have established that

Proposition 4.2.9 *The limit of the discounted approximation verifies $u_1 \in \mathcal{F}$.*

We continue this paragraph by establishing an alternative formula for the limit function u_1.

Proposition 4.2.10 *For all $x \in X$, the following equality holds:*

$$u_1(x) = \min_{\mu \in \widehat{\mathcal{P}}_0} \int_X h(y, x) \mathrm{d}\pi_{1*}\mu(y),$$

where u_1 is the function of Theorem 4.2.1 and h the Peierls barrier.

Proof We denote $\hat{u}$ the right hand side. We first claim that $\hat{u}$ is a subsolution. Indeed, each function $h_y = h(y, \cdot)$ is a subsolution by Proposition 3.1.2. Hence, if m is a Borel probability measure on X, so is h_m defined by $h_m(x) = \int_X h(y, x)\mathrm{d}m(y)$ since $\mathcal{S}$ is closed and convex (see Proposition 2.2.6). Finally, as $\hat{u}$ is an infimum of functions of this type, it is itself a subsolution by Lemma 3.2.6.

Next, we establish that $u_1 \leqslant \hat{u}$. Let $u \in \mathcal{S}$ be a continuous subsolution, we know that $u(x) - u(y) \leqslant h(y, x)$ for all pairs (x, y) (Proposition 3.1.2). Let $\mu \in \widehat{\mathcal{P}}_0$ be a Mather measure, integrating with respect to y the previous inequality yields

$$u(x) - \int_X u(y)\mathrm{d}\pi_{1*}\mu(y) \leqslant \int_X h(y, x)\mathrm{d}\pi_{1*}\mu(y).$$

If $u \in \mathcal{F}$ then we conclude that $u(x) \leqslant \int_X h(y, x)\mathrm{d}\pi_{1*}\mu(y)$. This being valid for all $u \in \mathcal{F}$ and for all $\mu \in \widehat{\mathcal{P}}_0$ we obtain the desired inequality $u_1 \leqslant \hat{u}$.

We conclude by proving the reverse inequality. Let $y \in X$, the function $h^y = -h(\cdot, y)$ is a subsolution (by Proposition 3.1.2). Moreover, by definition of $\hat{u}$, the function $h^y + \hat{u}(y) \in \mathcal{F}$. In particular, $u_1 \geqslant h^y + \hat{u}(y)$ and evaluating at y we obtain $u_1(y) \geqslant -h(y, y) + \hat{u}(y)$. If we specify moreover $y \in \mathcal{A}$ to be in the projected Aubry set then we have proved that (see Theorem 3.1.5):

$$\forall y \in \mathcal{A}, \quad u_1(y) \geqslant \hat{u}(y).$$

This is enough to conclude that $u_1 \geqslant \hat{u}$ everywhere, indeed, u_1 is a weak KAM solution and $\hat{u} \in \mathcal{S}$ hence Theorem 3.2.4 applies. □

Remark 4.2.11 The limit of the family $(u_\lambda)_{\lambda\in(0,1)}$ as $\lambda \to 1$ can be reformulated in terms of Mather measures on $X^{\mathbb{Z}}$. Indeed, as marginals of such measures are the same as those on $X \times X$ one finds that

$$\mathcal{F} = \left\{ u \in \mathcal{S} \cap C^0(X, \mathbb{R}), \forall \tilde{\mu} \in \widetilde{\mathcal{P}}_0, \quad \int_{X^{\mathbb{Z}}} u(x_0) \mathrm{d}\tilde{\mu}\big((x_n)_{n\in\mathbb{Z}}\big) \leqslant 0 \right\}.$$

And also, for all $x \in X$, the following equality holds:

$$u_1(x) = \min_{\tilde{\mu}\in\widetilde{\mathcal{P}}_0} \int_X h(x_0, x) \mathrm{d}\tilde{\mu}\big((x_n)_{n\in\mathbb{Z}}\big).$$

And finally, here is a mild property of u_1:

Proposition 4.2.12 *There exists a Mather measure $\mu_0 \in \widehat{\mathcal{P}}_0$ such that*

$$\int_X u_1(x) \mathrm{d}\pi_{1*}\mu_0(x) = 0.$$

Moreover, it can be imposed that μ_0 is an extremal point of $\widehat{\mathcal{P}}_0$.

Proof By Proposition 4.2.4 the selected function verifies $u_1 \in \mathcal{F}$ meaning that $\int_X u_1(x)\mathrm{d}\pi_{1*}\mu(x) \leqslant 0$ for all Mather measures $\mu \in \widehat{\mathcal{P}}_0$. If the result were not true, by compactness of $\widehat{\mathcal{P}}_0$ there would be an $\varepsilon > 0$ such that $\int_X u_1(x)\mathrm{d}\pi_{1*}\mu(x) \leqslant -\varepsilon$ for all Mather measures $\mu \in \widehat{\mathcal{P}}_0$. Then the function $u_1 + \varepsilon$ would also belong to $\mathcal{F}$ contradicting the definition of u_1 given in 4.2.5.

The second assertion is a direct consequence of Choquet's Theorem [182]. Indeed, it states that if μ_0 is a measure given by the first part of the Proposition, then there exists a probability measure w on $\widehat{\mathcal{P}}_0$, supported on the extremal points of $\mathcal{P}_0$ such that $\mu_0 = \int_{\widehat{\mathcal{P}}_0} \mu \mathrm{d}w(\mu)$. Any measure μ_1 in the support of w has to verify $\int_X u_1(x)\mathrm{d}\pi_{1*}\mu_1(x) = 0$. □

Remark 4.2.13 The previous Proposition holds as well when considering Mather measures as measures on $X^{\mathbb{Z}}$ thanks to the point of view of Bernard and Buffoni (see Remark 4.1.2). In this case, denoting by $\widetilde{\mathcal{P}}_0$ the set of minimizing shift invariant measures, extremal measures are the ergodic ones with respect to the action of the shift.

Before turning to the positive counterpart of those results let us provide a simplistic economical interpretation. As previously, X is the metric space of wine stores in France, $c : X \times X \to \mathbb{R}$ the cost of a 24 hour delivery and $R : X \to \mathbb{R}$ provides the price $R(x)$ of a bottle of Château Rayas[4] in the store x. The discount factor plays the role of an interest rate, or of inflation depending on the point of view.

[4] Château Rayas is definitively the best red wine, and arguably the best white wine, that the author has had the privilege of tasting. They are both of the appellation Châteauneuf du Pape which is the

If some money $m > 0$ is placed in the bank at a daily rate λ^{-1}, then tomorrow it will be worth $\lambda^{-1}m$. Conversely, if one buys today a bottle of Château Rayas at the price $R(y)$ but only pays it tomorrow, it is considered that the actualized price is $\lambda R(y)$ (as this amount of money put in the bank today will buy the bottle tomorrow at price $R(y)$). Henceforth taking into account this effect of time, the actualized least price to obtain a bottle of Château Rayas at x tomorrow is $T_\lambda^- R(x) = \inf_{y\in X} \lambda R(y) + c(y, x)$, considering that the transportation will be paid tomorrow at tomorrow's price.

In this context, the function u_λ, fixed point of T_λ^- is called equilibrium state. It is the only price function such that a buyer has not to worry about the time at which he wishes to receive his bottle. It is also the asymptotic price of a bottle for someone willing to wait a very long time, when the interest rate is at λ.

4.3 Discount for the Positive Lax–Oleinik Semigroup

We here address the positive counterpart of the previous results and explore some relations between the obtained limits. This is new to our knowledge.

Of course, all the constructions and results of the previous section hold for the positive Lax–Oleinik semigroup. If $\lambda \in [0, 1)$ we denote by v_λ the unique fixed point of the operator $T_\lambda^+ : u \mapsto T^+(\lambda u)$ that is a contraction. Similar arguments as in the previous paragraph yield:

Theorem 4.3.1 *There exists a positive weak KAM solution v_1 such that $v_\lambda - \frac{c[0]}{1-\lambda} \to v_1$ where the convergence takes place as $\lambda \to 1$ and is uniform.*

The functions v_λ have the following explicit form:

Lemma 4.3.2 *For any $\lambda \in (0, 1)$ and $x \in X$, we have*

$$v_\lambda(x) = -\min_{\substack{(x_n)_{n\geqslant 0}\\ x_0=x}} \sum_{n\geqslant 0} \lambda^n c(x_n, x_{n+1}).$$

The limit v_1 has the following form:

Proposition 4.3.3 *Let $\mathcal{F}^+ \subset \mathcal{S} \cap C^0(X, \mathbb{R})$ be the set of continuous subsolutions u verifying the constraint $\int_X u(x)\mathrm{d}\pi_{1*}\mu(x) \geqslant 0$ for all Mather measures $\mu \in \widehat{\mathcal{P}}_0$.*

We have the formulas $v_1 = \inf_{u\in\mathcal{F}^+} u$ where the infimum is taken pointwise.

The function v_1 verifies $v_1 \in \mathcal{F}^+$.

---

most prestigious of the meridional Rhône valley. Wines made by their owner, Emmanuel Reynaud, have no equal.

And finally for all $x \in X$,

$$v_1(x) = \max_{\mu \in \widehat{\mathcal{P}}_0} \int_X -h(x, y) \mathrm{d}\pi_{1*}\mu(y).$$

As for the negative Lax–Oleinik semigroup (Remark 4.2.11) the previous proposition can be stated in terms of Mather measures on $X^{\mathbb{Z}}$, which we leave to the reader.

We conclude by asking the following:

Question
what are the links between u_1 and v_1?

Unfortunately, the answer may seem disappointing, there is, in general no particular link. For example, except in very particular instances, they are not a conjugate pair (as they do not have any reason to coincide on the Mather set $\mathcal{M}$). They are not even ordered even though the following inequalities hold on the projected Aubry set:

Proposition 4.3.4 *The functions u_1 and v_1 verify*

$$\forall x \in \mathcal{A}, \quad u_1(x) \leqslant v_1(x).$$

Proof Let us argue by contradiction assuming that there exists $x_0 \in \mathcal{A}$ such that $v_1(x_0) < u_1(x_0)$. We set $\varepsilon = u_1(x_0) - v_1(x_0) > 0$. We will construct a Mather measure μ_0 such that $\int_X v_1(z) \mathrm{d}\pi_{1*}\mu_0(z) < 0$. This will be our contradiction as $v_1 \in \mathcal{F}^+$ meaning that $\int_X v_1(z) \mathrm{d}\pi_{1*}\mu_0(z) \geqslant 0$.

Let $(x_n)_{n\in\mathbb{Z}} \in \widetilde{\mathcal{A}}$ be a sequence associated to x_0. As u_1 and v_1 are critical subsolutions, one infers (see Remark 2.4.8) that

$$\forall n \geqslant 0, \quad u_1(x_0) - u_1(x_{-n}) = \sum_{k=-n}^{-1} c(x_k, x_{k+1}) + nc[0],$$

$$v_1(x_0) - v_1(x_{-n}) = \sum_{k=-n}^{-1} c(x_k, x_{k+1}) + nc[0].$$

It follows that $\varepsilon = u_1(x_{-n}) - v_1(x_{-n})$ for all $n \geqslant 0$. By continuity, one finds that $u_1 - v_1$ is constantly equal to ε on $\overline{\{x_{-n},\ n \geqslant 0\}}$.

Last, arguing as in the proof of Theorem 4.1.6, we construct a minimizing Mather measure $\mu_0 \in \widehat{\mathcal{P}}_0$ such that the support of $\pi_{1*}\mu_0$ is included in $\overline{\{x_{-n},\ n \geqslant 0\}}$. We conclude, using again that $u_1 \in \mathcal{F}$, that

$$\int_X v_1(z)\mathrm{d}\pi_{1*}\mu_0(z) \leqslant \int_X \big(v_1(z) - u_1(z)\big)\mathrm{d}\pi_{1*}\mu_0(z) = -\varepsilon.$$

□

The concluding general result here gives a condition for u_1 and v_1 to be a conjugate pair:

Proposition 4.3.5 *The following assertions are equivalent:*

1. *The functions u_1 and v_1 form a conjugate pair,*
2. $u_{1|\mathcal{A}} = v_{1|\mathcal{A}}$,
3. $u_1 \geqslant v_1$,
4. *for all Mather measures μ, the equality $\int_X u_1(x)\mathrm{d}\pi_{1*}\mu(x) = 0$ holds,*
5. *for all Mather measures μ, the equality $\int_X v_1(x)\mathrm{d}\pi_{1*}\mu(x) = 0$ holds,*
6. *there exists a critical subsolution $v \in \mathcal{S}$ such that for all Mather measures μ, the equality $\int_X v(x)\mathrm{d}\pi_{1*}\mu(x) = 0$ holds.*

Proof Assertion (1) being equivalent to (2) follows from the definition of a conjugate pair as explained in Remark 3.1.4.

If (2) holds, then (3) holds as this inequality is always true for a conjugate pair. Reciprocally, if (3) holds, then by Proposition 4.3.4, (2) is true.

Assertion (3) implies (4) and (5). It is an immediate consequence of the fact that $u_1 \in \mathcal{F}$ and $v_1 \in \mathcal{F}^+$.

Then, (4) or (5) implies (6) is straightforward as negative or positive weak KAM solutions are subsolutions.

Let us now establish (6) implies (2). Let v be the subsolution given by the hypothesis and let us denote by v^- and v^+ the respective limits of $T^{-n}v + nc[0]$ and $T^{+n}v - nc[0]$ as $n \to +\infty$. As $v^-_{|\mathcal{A}} = v^+_{|\mathcal{A}} = v_{|\mathcal{A}}$ we obtain respectively a negative and positive weak KAM solution satisfying the hypothesis of (6). The idea of the proof is that there can be at most one such negative weak KAM solution (and similarly, at most one such positive weak KAM solution).

To this aim, let $\mu \in \widehat{\mathcal{P}}_0$ so that

$$\int_X v^-(x)\mathrm{d}\pi_{1*}\mu(x) = \int_X v^+(x)\mathrm{d}\pi_{1*}\mu(x) = \int_X v(x)\mathrm{d}\pi_{1*}\mu(x) = 0.$$

As $v^- \in \mathcal{F}$, $v^- \leqslant u_1$ and as $u_1 \in \mathcal{F}$ it follows that

$$\forall \mu \in \widehat{\mathcal{P}}_0, \quad 0 = \int_X v^-(x)\mathrm{d}\pi_{1*}\mu(x) \leqslant \int_X u_1(x)\mathrm{d}\pi_{1*}\mu(x) \leqslant 0.$$

So u_1 itself satisfies the hypothesis of (5). Moreover, combining the previous equalities with $v^- \leqslant u_1$ implies that v^- and u_1 coincide on the support of μ (as both functions are continuous). This being true for all minimizing Mather measures $\mu \in \widehat{\mathcal{P}}_0$, we conclude that $u_{1|\mathcal{M}} = v^-_{|\mathcal{M}}$, by Theorem 4.1.6, we deduce that $u_1 = v^-$. The same proof yields that $v_1 = v^+$. Hence the pair (u_1, v_1) is a conjugate pair. □

4.4 Degenerate Discounted Equations

As an original contribution, let us finish by a generalization of the discounted convergence results. Instead of modifying the Lax–Oleinik semigroup to make it a contraction, we perturb it so that it is still a 1–Lipschitz map. Yet conditions are given in order to select again a weak KAM solution as the perturbation gets smaller. In this generality, the results of this paragraph are new.[5]

We consider a continuous function $\alpha : X \to \mathbb{R}$ that verifies the following two conditions:

(α1) the function α has values in $[0, 1)$,
(α2) for all minimizing Mather measure $\mu \in \widehat{\mathcal{P}}_0$, $\int_X \alpha(x) \mathrm{d}\pi_{1*}\mu(x) > 0$.

This last property is obviously verified if α is positive on the projected Aubry set $\mathcal{A}$ (this was the condition of [205]) or if α is positive on $\mathcal{M}$. The problem to be studied is understanding the behavior of functions $u_\lambda : X \to \mathbb{R}$, for $\lambda \in (0, 1)$, verifying

$$\forall x \in X, \quad u_\lambda(x) = T^-\big((1-\lambda\alpha)u_\lambda\big)(x) + c[0],$$

as $\lambda \to 0$. The convergence result is stated later on in this section in Theorem 4.4.12. Therefore, let us denote by $\mathfrak{T}_\lambda$ the mapping $v \mapsto T^-\big((1-\lambda\alpha)u_\lambda\big) + c[0]$. Just like T^- (see Proposition 2.1.3), the operators $\mathfrak{T}_\lambda$ are 1–Lipschitz and order preserving.

We start by easy properties in order to get acquainted with the operators:

Proposition 4.4.1 *Let $v : X \to \mathbb{R}$ be a continuous function, then*

$$\forall x \in X, \exists x_{-1} \in X, \quad \mathfrak{T}_\lambda v(x) = \big(1 - \lambda\alpha(x_{-1})\big)v(x_{-1}) + c(x_{-1}, x) + c[0].$$

[5] This has since been extended in [180].

More generally, for all $n > 0$*, there is a chain* $(x_{-n}, \cdots, x_0 = x)$ *such that*

$$\mathfrak{T}_\lambda^{-n} v(x) = \beta_{-n} v(x_{-n}) + \sum_{k=-n}^{-1} \beta_{k+1}\big(c(x_k, x_{k+1}) + c[0]\big),$$

where $\beta_k = \prod_{j=k}^{-1} \big(1 - \lambda\alpha(x_j)\big)$*, for* $-n \leqslant k \leqslant -1$ *and* $\beta_0 = 1$.

Proof The first point is a direct consequence of compactness and continuity while the second follows from a straightforward induction. □

Beware that the notation β_k is misleading as it depends on the chain $(x_k, \cdots, x_0)$. We now address the issue of fixed points of $\mathfrak{T}_\lambda$.

Definition 4.4.2 We will say a function $u : X \to \mathbb{R}$ is a λ–discounted subsolution if $u \leqslant \mathfrak{T}_\lambda u$ or equivalently

$$\forall (x, y) \in X \times X, \quad u(x) - \big(1 - \lambda\alpha(y)\big)u(y) \leqslant c(y, x) + c[0]. \tag{4.1}$$

A function $v : X \to \mathbb{R}$ is a λ–discounted solution if $v = \mathfrak{T}_\lambda v$.

By definition and successive applications of Proposition 4.4.1 one gets:

Proposition 4.4.3 *Let* $u : X \to \mathbb{R}$ *be a λ–discounted subsolution, then for all* $n > 0$ *and all finite chains* $(y_{-n}, \cdots, y_0 = x)$*,*

$$u(y_0) \leqslant \beta_{-n} u(y_{-n}) + \sum_{k=-n}^{-1} \beta_{k+1}\big(c(y_k, y_{k+1}) + c[0]\big),$$

where $\beta_k = \prod_{j=k}^{-1} \big(1 - \lambda\alpha(y_j)\big)$ *and* $\beta_0 = 1$.

Let $v : X \to \mathbb{R}$ *be a λ–discounted solution. Then for all* $x \in X$*, there exists an infinite chain* $(x_k)_{k\leqslant 0}$ *such that* $x_0 = x$ *and*

$$\forall n > 0, \quad v(x) = \beta_{-n} v(x_{-n}) + \sum_{k=-n}^{-1} \beta_{k+1}\big(c(x_k, x_{k+1}) + c[0]\big),$$

where $\beta_k = \prod_{j=k}^{-1} \big(1 - \lambda\alpha(x_j)\big)$ *and* $\beta_0 = 1$.

The convention adopted here is that an empty product has value 1, so that in the previous notation, the formula also holds for β_0.

The next result is reminiscent of strong comparison principles in viscosity solutions theory:

Theorem 4.4.4 *Let $\lambda \in (0,1)$, $u : X \to \mathbb{R}$ be a λ–discounted subsolution and $v : X \to \mathbb{R}$ be a λ–discounted solution. Then $u \leqslant v$.*

Proof As $u \leqslant \mathfrak{T}_\lambda u$, it is enough to prove that $\mathfrak{T}_\lambda u \leqslant v$. Hence by Proposition 2.1.3, one assumes that u is continuous, without loss of generality. Then consider a strict subsolution $u_0 : X \to \mathbb{R}$ given by Theorem 2.4.1. Moreover, up to subtracting a big constant, we assume that u_0 is negative. For $\varepsilon \in (0,1)$ we define $u_\varepsilon = (1-\varepsilon)u_0 + \varepsilon u$. The function u_ε is a λ–discounted subsolution. As a matter of fact, if $(x, y) \in X \times X$,

$$\begin{aligned}
&u_\varepsilon(x) - \big(1 - \lambda\alpha(y)\big)u_\varepsilon(y) \\
&\quad = \varepsilon\big(u(x) - \big(1 - \lambda\alpha(y)\big)u(y)\big) + (1-\varepsilon)\big(u_0(x) - \big(1 - \lambda\alpha(y)\big)u_0(y)\big) \\
&\quad \leqslant \varepsilon\big(c(y,x) + c[0]\big) + (1-\varepsilon)\big(u_0(x) - u_0(y)\big) \leqslant c(y,x) + c[0],
\end{aligned} \tag{4.2}$$

where it was used first that u_0 is negative and then that it is a critical subsolution.

Let now $x_0 \in X$ such that $u_\varepsilon(x_0) - v(x_0) = \max(u_\varepsilon - v)$. We aim at proving that $u_\varepsilon(x_0) - v(x_0) \leqslant 0$. Let us argue by contradiction, assuming that $u_\varepsilon(x_0) - v(x_0) > 0$. Let $(x_k)_{k\leqslant 0}$ be a chain given by Proposition 4.4.3 for v, $(\beta_k)_{k\leqslant 0}$ the associated sequence as defined in the same Proposition 4.4.3. It follows from both assertions of Proposition 4.4.3 that for all $k < 0$,

$$\begin{aligned}
(u_\varepsilon - v)(x_0) &= u_\varepsilon(x_0) - \beta_k v(x_k) - \sum_{j=k}^{-1} \beta_{j+1}\big(c(x_j, x_{j+1}) + c[0]\big) \\
&\leqslant \beta_k u_\varepsilon(x_k) + \sum_{j=k}^{-1} \beta_{j+1}\big(c(x_j, x_{j+1}) + c[0]\big) \\
&\quad - \beta_k v(x_k) - \sum_{j=k}^{-1} \beta_{j+1}\big(c(x_j, x_{j+1}) + c[0]\big) \\
&= \beta_k(u_\varepsilon - v)(x_k) \leqslant (u_\varepsilon - v)(x_k),
\end{aligned}$$

where the last inequality is obtained using the contradiction hypothesis and the inequalities $0 < \beta_k \leqslant 1$. By definition of x_0, it follows that all the preceding inequalities are equalities. In particular, it comes that $\beta_k = 1$ for all $k < 0$ which in turn implies that $\alpha(x_k) = 0$ for all $k \leqslant 0$, by definition of β_k. Moreover, tracing the inequalities used, it follows that

$$\forall k \leqslant 0, \quad u_\varepsilon(x_0) = \beta_k u_\varepsilon(x_k) + \sum_{j=k}^{-1} \beta_{j+1}\big(c(x_j, x_{j+1}) + c[0]\big).$$

Going back to (4.2) and using that, there as well, inequalities are indeed equalities, it follows that $u_0(x_k) - u_0(x_{k+1}) = c(x_k, x_{k+1}) + c[0]$ for all $k < 0$. By definition of u_0 and thanks to its property of being strict, we conclude that $(x_k, x_{k+1}) \in \widehat{\mathcal{A}}$ for all $k < 0$.

Let us now define, for $n > 0$ the probability measure $\mu_n = \frac{1}{n} \sum_{k=-n}^{-1} \delta_{(x_k, x_{k+1})}$ (that is supported on $\widehat{\mathcal{A}}$). Let μ be an accumulation point of the sequence of probability measures $(\mu_n)_{n>0}$ for some subsequence $(n_i)_{i \geqslant 0}$. Arguing as in the proof of Theorem 4.1.3, we find that the measure μ is closed. As the 2-Aubry set is closed, the measure μ is supported on $\widehat{\mathcal{A}}$. The last part of Theorem 4.1.3 implies that μ is a Mather minimizing measure.

Finally, using that $\alpha(x_k) = 0$ for all $k < 0$ observe that

$$\int_X \alpha(x) \mathrm{d}\pi_{1*}\mu(x) = \lim_{i \to +\infty} \int_X \alpha(x) \mathrm{d}\pi_{1*}\mu_{n_i}(x) = \lim_{i \to +\infty} \frac{1}{n_i} \sum_{k=-n_i}^{-1} \alpha(x_k) = 0,$$

thus contradicting Hypothesis (α2). Hence $u_\varepsilon(x_0) - v(x_0) = \max(u_\varepsilon - v) \leqslant 0$ and $u_\varepsilon \leqslant v$. As this holds for all $\varepsilon \in (0, 1)$, letting $\varepsilon \to 1$ proves that $u \leqslant v$. □

The previous proof combines two main ideas. The first one is that subsolutions can be approximated by strict subsolutions, thus forcing interesting phenomena to take place on the Aubry set. This is made possible by the convex structure of our minimization problems. The second idea is to construct illicit Mather measures assuming that subsolutions or solutions do not verify suitable properties. This line of reasoning will be used several times in what follows.

As λ–discounted solutions are obviously λ–discounted subsolutions, the previous Proposition brings as a consequence that there can be at most one λ–discounted solution. The next existence result shows there is exactly one:

Theorem 4.4.5 *For all $\lambda \in (0, 1)$ there exists a unique λ–discounted solution.*

Proof Let $\underline{u} < 0$ be a negative weak KAM solution and $\overline{u} > 0$ be a positive weak KAM solution. Applying the modified Lax–Oleinik semigroup yields

$$\mathfrak{T}_\lambda(\underline{u}) = T^-\big((1 - \lambda\alpha)\underline{u})\big) + c[0] \geqslant T^-(\underline{u}) + c[0] = \underline{u}.$$

A straightforward induction yields that the sequence $\big(\mathfrak{T}_\lambda^n(\underline{u})\big)_{n \geqslant 0}$ is non–decreasing.

Similarly,

$$\mathfrak{T}_\lambda(\overline{u}) = T^-\big((1 - \lambda\alpha)\overline{u})\big) + c[0] \leqslant T^-(\overline{u}) + c[0] = \overline{u}.$$

A straightforward induction yields that the sequence $\big(\mathfrak{T}_\lambda^n(\overline{u})\big)_{n \geqslant 0}$ is non–increasing.

Finally, as $\underline{u} < \overline{u}$ it follows that $\mathfrak{T}_\lambda^n(\underline{u}) \leqslant \mathfrak{T}_\lambda^n(\overline{u})$ for all $n \geqslant 0$. The sequence $\big(\mathfrak{T}_\lambda^n(\underline{u})\big)_{n \geqslant 0}$ is bounded and non–decreasing, made of equi–continuous functions, hence it converges (uniformly) towards a function $u_\lambda : X \to \mathbb{R}$, verifying $\underline{u} \leqslant u_\lambda \leqslant \overline{u}$, that is, by continuity of $\mathfrak{T}_\lambda$, a λ–discounted solution.

□

Definition 4.4.6 For all $\lambda \in (0, 1)$, the unique λ–discounted solution is denoted by u_λ^α.

As a byproduct of the previous proof, it was established:

Corollary 4.4.7 *The family $(u_\lambda^\alpha)_{\lambda \in (0,1)}$ is uniformly bounded and consists of equi–continuous functions.*

The last part holds as the u_λ^α are in the image of T^- (Proposition 2.1.3). As the family $(u_\lambda^\alpha)_{\lambda \in (0,1)}$ is relatively compact, to prove that it converges when $\lambda \to 0$, it is enough to prove there is a unique accumulation point.

The next proposition establishes the crucial property of such accumulation points, similarly to Proposition 4.2.4:

Proposition 4.4.8 *Let $\mu \in \widehat{\mathcal{P}}_0$ be a Mather measure. Assume $u_{\lambda_n}^\alpha \to u$ as $n \to +\infty$ for some extraction $\lambda_n \to 0$. Then $\int_X \alpha(x) u(x)\, \mathrm{d}\pi_{1*}\mu(x) \leqslant 0$.*

Proof Let us start from the family of inequalities given by (4.1), applied to the functions u_λ. Integrating against μ it is obtained that

$$0 = \int_{X\times X} \big[c(y,x) + c[0]\big] \mathrm{d}\mu(y,x) \geqslant \int_{X\times X} \big[u_\lambda^\alpha(x) - \big(1 - \lambda\alpha(y)\big) u_\lambda^\alpha(y)\big]\, \mathrm{d}\mu(y,x).$$

As μ is closed and u_λ continuous, dividing by λ, we gather that

$$\forall \lambda \in (0,1), \qquad \int_X \alpha(y) u_\lambda^\alpha(y)\, \mathrm{d}\pi_{1*}\mu(y) \leqslant 0.$$

Passing to the limit along the subsequence $(\lambda_n)_{n \geqslant 0}$, yields the result. □

Particular Mather measures can then be constructed starting from calibrating chains given by Proposition 4.4.3. One first needs to establish a crucial property they satisfy:

Proposition 4.4.9 *There exists $M > 0$ such that for all $\lambda \in (0, 1)$ and $x_0 \in X$, if $(x_k^\lambda)_{k \leqslant 0}$ is a sequence given by Proposition 4.4.3 applied to u_λ^α with $x_0^\lambda = x_0$, then*

$$\lambda \sum_{k \leqslant 0} \prod_{j=k}^{-1} \big(1 - \lambda \alpha(x_j^\lambda)\big) < M.$$

Proof Let us argue by contradiction assuming the result does not hold. Then there exist a sequence $(\lambda_n)_{n \in \mathbb{N}} \in (0,1)^{\mathbb{N}}$ and points $x_0^n \in X$ such that for each integer $n \in \mathbb{N}$ there exists a sequence $(x_k^n)_{k \leqslant 0}$ given by Proposition 4.4.3 associated to $u_{\lambda_n}^\alpha$

and an integer $N_n > 0$ such that $\lambda_n C_n = \lambda_n \sum_{k=-N_n}^{-1} \beta_{k+1}^n \to +\infty$, having adopted the notation $\beta_k^n = \prod_{j=k}^{-1} \big(1 - \lambda_n \alpha(x_j^n)\big)$. This implies that $N_n \to +\infty$ as $0 < \beta_k^n \leqslant 1$.

For all integer $n \in \mathbb{N}$, let us define the probability measure on $X \times X$,

$$\mu_n = C_n^{-1} \sum_{k=-N_n}^{-1} \beta_{k+1}^n \delta_{(x_k^n, x_{k+1}^n)}.$$

Up to an extraction, let us assume furthermore that the sequence μ_n converges to a probability measure μ. We will prove that μ is a minimizing Mather measure violating condition (α2).

The measure μ is closed: let $f : X \to \mathbb{R}$ be a continuous function. We compute, using an Abel transform:

$$\Big| \int_{X\times X} \big(f(y) - f(x)\big) \mathrm{d}\mu_n(x, y)\Big| = C_n^{-1} \Big| \sum_{k=-N_n}^{-1} \beta_{k+1}^n \big(f(x_{k+1}^n) - f(x_k^n)\big)\Big|$$

$$= C_n^{-1} \Big| \sum_{k=-N_n}^{-1} (\beta_k^n - \beta_{k+1}^n) f(x_k^n) - \beta_{-N_n}^n f(x_{-N_n}^n) + \beta_0 f(x_0^n)\Big|$$

$$\leqslant C_n^{-1} \Big[\sum_{k=-N_n}^{-1} (\beta_{k+1}^n - \beta_k^n) \|f\|_\infty + 2\|f\|_\infty \Big] \leqslant 4 C_n^{-1} \|f\|_\infty.$$

In the previous chain of inequalities it was used that the sequences $(\beta_k^n)_{k\leqslant 0}$ are non–decreasing and take values in $[0, 1]$. As $C_n \to +\infty$, letting $n \to +\infty$, it is obtained that

$$\int_{X\times X} \big(f(y) - f(x)\big) \mathrm{d}\mu(x, y) = \lim_{n\to+\infty} \int_{X\times X} \big(f(y) - f(x)\big) \mathrm{d}\mu_n(x, y) = 0.$$

Therefore μ is closed.

The measure μ is minimizing: we use the definition of μ_n and the property of the sequences $(x_k^n)_{k\leqslant 0}$.

$$\Big| \int_{X\times X} \big(c(x, y) + c[0]\big) \mathrm{d}\mu_n(x, y)\Big| = C_n^{-1} \Big| \sum_{k=-N_n}^{-1} \beta_{k+1}^n \big(c(x_k^n, x_{k+1}^n) + c[0]\big)\Big|$$

$$= C_n^{-1} \Big| \big(u_{\lambda_n}^\alpha(x_0^n) - \beta_{-N_n}^n u_{\lambda_n}^\alpha(x_{-N_n}^n)\big)\Big|$$

$$\leqslant 2 C_n^{-1} \|u_{\lambda_n}^\alpha\|_\infty.$$

Recalling that the family $(u_\lambda)_{\lambda^\alpha \in (0,1)}$ is uniformly bounded (Corollary 4.4.7) letting $n \to +\infty$ it follows that $\int_{X \times X} c(x, y)\, d\mu(x, y) = -c[0]$.

The measure μ satisfies $\int \alpha\, d\mu = 0$: we use the inequality $\exp(x) \geqslant 1+x$ and the definition of β_k^n to estimate

$$\begin{aligned}
\int_{X \times X} \alpha(x)\, d\mu_n(x, y) &= C_n^{-1} \sum_{k=-N_n}^{-1} \beta_{k+1}^n \alpha(x_k^n) \\
&\leqslant C_n^{-1} \sum_{k=-N_n}^{-1} \alpha(x_k^n) \exp\Big(- \lambda_n \sum_{j=k+1}^{-1} \alpha(x_j^n)\Big) \\
&\leqslant \frac{\exp(\|\alpha\|_\infty)}{C_n} \sum_{k=-N_n}^{-1} \alpha(x_k^n) \exp\Big(- \lambda_n \sum_{j=k}^{-1} \alpha(x_j^n)\Big).
\end{aligned}$$

As the α_k^n are non–negative and the function $x \mapsto \exp(-x)$ is decreasing, the right hand side can be estimated by comparing sum and integral to conclude that

$$\int_{X \times X} \alpha(x)\, d\mu_n(x, y) \leqslant \frac{\exp(\|\alpha\|_\infty)}{C_n} \int_0^\infty \exp(-\lambda_n x)\, dx = \frac{\exp(\|\alpha\|_\infty)}{\lambda_n C_n}.$$

As $\lambda_n C_n \to +\infty$, it follows that $\int_{X \times X} \alpha(x)\, d\mu(x, y) = 0$. Thus μ is a Mather measure contradicting $(\alpha 2)$ and the result is proved. □

As a Corollary, a refined representation formula comes up for the functions u_λ^α:

Corollary 4.4.10 *Let $\lambda \in (0, 1)$ and $x_0 \in X$. If $(x_k)_{k \leqslant 0}$ is given by Proposition 4.4.3 applied to u_λ^α, then*

$$u_\lambda^\alpha(x_0) = \sum_{k=-\infty}^{-1} \beta_{k+1}\big(c(x_k, x_{k+1}) + c[0]\big),$$

with $\beta_k = \prod_{j=k}^{-1} \big(1 - \lambda\alpha(x_j)\big)$ and $\beta_0 = 1$.

Proof By Proposition 4.4.9, the sum $\sum \beta_k$ is convergent which implies that $\lim_{k \to -\infty} \beta_k = 0$. As the function u_λ^α is bounded, the result follows by simply letting $n \to +\infty$ in the second part of Proposition 4.4.3. □

Let us now enter the convergence part of this section. Motivated by Proposition 4.4.8 we give the following definition:

Definition 4.4.11 Let $\mathcal{F}_\alpha$ be the set of continuous critical subsolutions $u : X \to \mathbb{R}$ such that $\int_X \alpha(x) u(x)\, d\pi_{1*}\mu(x) \leqslant 0$ for all Mather measure $\mu \in \widehat{\mathcal{P}}_0$.

Let us first state the main Theorem. The careful reader will notice quite a resemblance with Theorem 4.2.1 and Proposition 4.2.10:

Theorem 4.4.12 *The family of functions $(u_\lambda^\alpha)_{(0,1)}$ uniformly converges as $\lambda \to 0$. Moreover, denoting by u_0^α the limit, the two following formulas hold:*

- *for $x_0 \in X$, $u_0^\alpha(x_0) = \max_{u \in \mathcal{F}_\alpha} u(x_0)$;*
- *for $x_0 \in X$,*

$$u_0^\alpha(x_0) = \min_{\mu \in \widehat{\mathcal{P}}_0} \frac{\int_X \alpha(x) h(x, x_0)\, \mathrm{d}\pi_{1*}\mu(x)}{\int_X \alpha(x)\, \mathrm{d}\pi_{1*}\mu(x)},$$

where $h : X \times X \to \mathbb{R}$ still denotes Peierls' barrier given by Definition 3.1.1.

The proof of this Theorem is split into several Lemmas resembling what was done for the standard discounted equation.

Definition 4.4.13 If $\lambda \in (0, 1)$ and $x_0 \in X$, we choose a sequence $(x_k^\lambda)_{k \leqslant 0}$ given by Proposition 4.4.3 applied to u_λ^α with $x_0 = x_0^\lambda$. The probability measure $\mu_{x_0}^\lambda$ is defined by:

$$\mu_{x_0}^\lambda = C_{x_0,\lambda}^{-1} \sum_{k=-\infty}^{-1} \beta_{k+1}^{x_0,\lambda} \delta_{(x_k^\lambda, x_{k+1}^\lambda)},$$

where $\beta_k^{x_0,\lambda} = \prod_{j=k}^{-1} \left(1 - \lambda\alpha(x_j^\lambda)\right)$ and $C_{x_0,\lambda} = \sum_{k=-\infty}^{-1} \beta_{k+1}^{x_0,\lambda}$.

The sum defining $C_{x_0,\lambda}$ is indeed finite by Proposition 4.4.9.

Lemma 4.4.14 *Let $x_0 \in X$ and $\lambda_n \to 0$ be a sequence such that the family of measures $(\mu_{x_0}^{\lambda_n})_{n \in \mathbb{N}}$ converges to a probability measure μ. Then μ is a minimizing Mather measure.*

Proof We first prove that $C_{x_0,\lambda_n} \to +\infty$. Indeed, for all $n > 0$ and $k \leqslant 0$, $\beta_k^{x_0,\lambda_n} \geqslant (1 - \lambda_n \|\alpha\|_\infty)^{|k|}$ thus implying that

$$C_{x_0,\lambda_n} \geqslant \sum_{j=0}^{+\infty} (1 - \lambda_n \|\alpha\|_\infty)^j = \frac{1}{\lambda_n \|\alpha\|_\infty} \underset{n \to +\infty}{\longrightarrow} +\infty.$$

By computations the reader should already be familiar with, from the proof of Proposition 4.4.9, it is proven that μ is closed. Let $f : X \to \mathbb{R}$ be a continuous function. We compute using an Abel transform:

$$\left| \int_{X \times X} (f(y) - f(x))\, \mathrm{d}\mu_{x_0}^{\lambda_n}(x, y) \right| = C_{x_0,\lambda_n}^{-1} \left| \sum_{k=-\infty}^{-1} \beta_{k+1}^{x_0,\lambda_n} \left(f(x_{k+1}^{\lambda_n}) - f(x_k^{\lambda_n})\right) \right|$$

$$= C_{x_0,\lambda_n}^{-1}\Big|\sum_{k=-\infty}^{-1}(\beta_k^{x_0,\lambda_n}-\beta_{k+1}^{x_0,\lambda_n})f(x_k^{x_0,\lambda_n})+\beta_0^{x_0,\lambda_n}f(x_0)\Big|$$

$$\leqslant C_{x_0,\lambda_n}^{-1}\Big[\sum_{k=-\infty}^{-1}(\beta_{k+1}^{x_0,\lambda_n}-\beta_k^{x_0,\lambda_n})\|f\|_\infty+\|f\|_\infty\Big]\leqslant 2C_{x_0,\lambda_n}^{-1}\|f\|_\infty.$$

As $C_{x_0,\lambda_n}\to+\infty$ this proves that $\int_{X\times X}\big(f(y)-f(x)\big)\,\mathrm{d}\mu(x,y)=0$.

And then it is established that μ is minimizing:

$$\Big|\int_{X\times X}\big(c(x,y)+c[0]\big)\mathrm{d}\mu_{x_0}^{\lambda_n}(x,y)\Big|$$

$$= C_{x_0,\lambda_n}^{-1}\Big|\sum_{k=-\infty}^{-1}\beta_{k+1}^{x_0,\lambda_n}\big(c(x_k^{\lambda_n},x_{k+1}^{\lambda_n})+c[0]\big)\Big|$$

$$= C_{x_0,\lambda_n}^{-1}\big|u_{\lambda_n}^{\alpha}(x_0)\big|\leqslant C_{x_0,\lambda_n}^{-1}\|u_{\lambda_n}^{\alpha}\|_\infty.$$

Corollary 4.4.10 was used for the last equality. Recalling that the family $(u_\lambda^\alpha)_{\lambda\in(0,1)}$ is uniformly bounded (Corollary 4.4.7), letting $n\to+\infty$ it follows that

$$\int_{X\times X}c(x,y)\,\mathrm{d}\mu(x,y)=-c[0],$$

thus concluding the proof.

□

The next lemma is similar to Lemma 4.2.8:

Lemma 4.4.15 *Let $x_0\in X$, $\lambda\in(0,1)$ and $w\in\mathcal{S}$ be a continuous subsolution, then*

$$u_\lambda^\alpha(x_0)\geqslant w(x_0)-\lambda C_{x_0,\lambda}\int_X\alpha(z)w(z)\,\mathrm{d}\pi_{1*}\mu_{x_0}^{\lambda}(z).$$

Proof We use Corollary 4.4.10 and the fact that w is a critical subsolution:

$$u_\lambda^\alpha(x_0)=\sum_{k=-\infty}^{-1}\beta_{k+1}^{x_0,\lambda}\big(c(x_k^\lambda,x_{k+1}^\lambda)+c[0]\big)$$

$$\geqslant\sum_{k=-\infty}^{-1}\beta_{k+1}^{x_0,\lambda}\big(w(x_{k+1}^\lambda)-w(x_k^\lambda)\big)$$

$$=\sum_{k=-\infty}^{-1}(\beta_k^{x_0,\lambda}-\beta_{k+1}^{x_0,\lambda})w(x_k^{x_0,\lambda})+\beta_0^{x_0,\lambda}w(x_0).$$

The last equality follows by an Abel transform. We now use the definition of $\beta_k^{x_0,\lambda_n}$ to compute

$$\beta_k^{x_0,\lambda} - \beta_{k+1}^{x_0,\lambda} = \prod_{j=k}^{-1}\left(1-\lambda\alpha(x_j^\lambda)\right) - \prod_{j=k+1}^{-1}\left(1-\lambda\alpha(x_j^\lambda)\right)$$

$$= -\lambda\alpha(x_k^\lambda)\prod_{j=k+1}^{-1}\left(1-\lambda\alpha(x_j^\lambda)\right) = -\lambda\alpha(x_k^\lambda)\beta_{k+1}^{x_0,\lambda}.$$

Going back to the previous computation and remembering that $\beta_0^{x_0,\lambda} = 1$ yields

$$u_\lambda^\alpha(x_0) \geqslant w(x_0) - \lambda\sum_{k=-\infty}^{-1}\beta_{k+1}^{x_0,\lambda} w(x_k^{x_0,\lambda})\alpha(x_k^\lambda)$$

$$= w(x_0) - \lambda C_{x_0,\lambda}\int_X \alpha(z)w(z)\,\mathrm{d}\pi_{1*}\mu_{x_0}^\lambda(z).$$

□

The first part of Theorem 4.4.12 is now ready to be proven:

Proof of Theorem 4.4.12 First Formula Let $\lambda_n \to 0$ be a sequence such that $(u_{\lambda_n}^\alpha)_{n\in\mathbb{N}}$ converges to a function $v : X \to \mathbb{R}$. Henceforth the function v is a weak KAM solution by continuity of the Lax–Oleinik operator. We have also defined for all $x \in X$, $u_0^\alpha(x) = \max\limits_{u\in\mathcal{F}_\alpha} u(x)$. The aim here is to prove that $v = u_0^\alpha$.

By Proposition 4.4.8, $v \in \mathcal{F}_\alpha$ and therefore $v \leqslant u_0^\alpha$.

Let us now prove the reverse inequality. Let $x_0 \in X$. Up to a further extraction, we assume that the sequence of probability measures $(\mu_{x_0}^{\lambda_n})_{n\in\mathbb{N}}$ weakly converges to a measure μ that is a minimizing Mather measure thanks to Lemma 4.4.14. If $w \in \mathcal{F}_\alpha$, by definition, $\int_X \alpha(x)w(x)\,\mathrm{d}\pi_{1*}\mu(x) \leqslant 0$. Combining with Proposition 4.4.9 entails that $\limsup\limits_{n\to+\infty}\lambda_n C_{x_0,\lambda_n}\int_X \alpha(z)w(z)\,\mathrm{d}\pi_{1*}\mu_{x_0}^{\lambda_n}(z) \leqslant 0$. Plugging into the inequality of Lemma 4.4.15 and letting $n \to +\infty$ gives

$$v(x_0) \geqslant w(x_0) - \limsup_{n\to+\infty}\lambda_n C_{x_0,\lambda_n}\int_X \alpha(z)w(z)\,\mathrm{d}\pi_{1*}\mu_{x_0}^{\lambda_n}(z) \geqslant w(x_0).$$

As this holds for all $w \in \mathcal{F}_\alpha$ it comes that $v(x_0) \geqslant u_0^\alpha(x_0)$ and being true for all $x_0 \in X$, the first convergence formula is proven. □

This section ends by establishing the second representation formula for u_0. To this aim, we set $\hat{u}_0^\alpha(x_0) = \min\limits_{\mu\in\widehat{\mathcal{P}}_0}\frac{\int_X \alpha(x)h(x,x_0)\mathrm{d}\pi_{1*}\mu(x)}{\int_X \alpha(x)\mathrm{d}\pi_{1*}\mu(x)}$ for all $x_0 \in X$, we will prove that $u_0^\alpha = \hat{u}_0^\alpha$. The proof follows closely that of Proposition 4.2.10:

Proof of Theorem 4.4.12 Second Formula We first claim that $\hat{u}_0^\alpha$ is a subsolution. Indeed, each function $h_y = h(y, \cdot)$ is a subsolution by Proposition 3.1.2. Hence, if μ is a probability measure on X, so is h_μ^α defined by $h_\mu^\alpha(x) = \frac{\int_X \alpha(y)h(y,x)\mathrm{d}\mu(y)}{\int_X \alpha(y)\mathrm{d}\mu(y)}$ since $\mathcal{S}$ is closed and convex (see Proposition 2.2.6). Last, as $\hat{u}_0^\alpha$ is an infimum of functions of this type, it is itself a subsolution by Lemma 3.2.6.

Next, we establish that $u_0^\alpha \leqslant \hat{u}_0^\alpha$. Let $u \in \mathcal{S}$ be a continuous subsolution, we know that $\alpha(y)\big(u(x) - u(y)\big) \leqslant \alpha(y)h(y, x)$ for all pairs (x, y) (Proposition 3.1.2). Let $\mu \in \widehat{\mathcal{P}}_0$ be a Mather measure, integrating with respect to y the previous inequality yields

$$\left(\int_X \alpha(y)\, \mathrm{d}\pi_{1*}\mu(y)\right) u(x) - \int_X \alpha(y)u(y)\, \mathrm{d}\pi_{1*}\mu(y) \leqslant \int_X \alpha(y)h(y, x)\, \mathrm{d}\pi_{1*}\mu(y).$$

If $u \in \mathcal{F}_\alpha$ it follows that $\left(\int_X \alpha(y)\, \mathrm{d}\pi_{1*}\mu(y)\right)u(x) \leqslant \int_X \alpha(y)h(y, x)\, \mathrm{d}\pi_{1*}\mu(y)$. This being valid for all $u \in \mathcal{F}_\alpha$ and for all $\mu \in \widehat{\mathcal{P}}_0$, the desired inequality $u_0^\alpha \leqslant \hat{u}_0^\alpha$ is obtained.

We conclude by proving the reverse inequality. Let $y \in X$, the function $h^y = -h(\cdot, y)$ is a subsolution (by Proposition 3.1.2). Moreover, by definition of $\hat{u}_0^\alpha$, the function $h^y + \hat{u}_0^\alpha(y) \in \mathcal{F}_\alpha$. In particular, $u_0^\alpha \geqslant h^y - \hat{u}_0^\alpha(y)$ and evaluating at y yields $u_0^\alpha(y) \geqslant -h(y, y) + \hat{u}_0^\alpha(y)$. If we specify, moreover, $y \in \mathcal{A}$ to be in the projected Aubry set, leads to the inequalities (see Theorem 3.1.5)

$$\forall y \in \mathcal{A}, \quad u_0^\alpha(y) \geqslant \hat{u}_0^\alpha(y).$$

This is enough to conclude that $u_0^\alpha \geqslant \hat{u}_0^\alpha$ everywhere, indeed, u_0^α is a weak KAM solution and $\hat{u}_0^\alpha \in \mathcal{S}$ hence Theorem 3.2.4 applies.

□

4.5 Comment on the Discounted Procedure

As already mentioned, the introduction of the functions u_λ in the second proof of the Weak KAM Theorem 2.2.1 is very natural. Indeed, let us recall a classical fixed point Theorem:

Theorem 4.5.1 *Let C be a compact convex subset of a Fréchet vector–space and $f : C \to C$ be a 1–Lipschitz map. Then f admits a fixed point.*

This result is of course weaker than the Schauder-Tychonoff Theorem but a simple proof goes as follows. Up to conjugating by a translation, assume that $0 \in C$. Then for $\lambda \in (0, 1)$, the function $f_\lambda : C \to C$ defined by $f_\lambda(x) = f(\lambda x)$ is well defined and a contraction of a complete metric space. It admits a unique fixed point $x_\lambda \in C$. Consequently, by compactness of C one can consider a sequence $\lambda_n \to 1$

such that $(x_{\lambda_n})_{n\in\mathbb{N}}$ converges to a point $x^* \in C$. It is then immediate that x^* is a fixed point of f. Note that if 0 is a fixed point of f, then $x_\lambda = 0$ for all $\lambda \in (0, 1)$.

A natural question is to figure out if in the previous procedure, the whole family x_λ always converges. If this were the case, our discounted Theorem 4.2.1 would be less interesting. However, this is not the case as we now illustrate.

Our example is constructed in $(\mathbb{R}^2, \|\cdot\|_1)$. More precisely, let us consider the triangle defined by

$$\mathfrak{T} = \left\{(x, y) \in \mathbb{R}^2,\ -\frac{1}{2} \leqslant y \leqslant -|x| + \frac{1}{2}\right\}.$$

If $\alpha \in (0, 1)$, we look for a map f that takes the following form:

$$f(x, y) = \left(x + \varepsilon(y), \alpha\left(y + \frac{1}{2}\right) - \frac{1}{2}\right),$$

where $\varepsilon : \left[-\frac{1}{2}, \frac{1}{2}\right] \to \mathbb{R}$ is a map to be determined such that $\varepsilon(-\frac{1}{2}) = 0$. In this setting, the bottom edge of $\mathfrak{T}$ is made of fixed points of f.

Simple verifications show that $f : \mathfrak{T} \to \mathfrak{T}$ is well defined as soon as $|\varepsilon(y)| \leqslant (1-\alpha)(y + \frac{1}{2})$. Moreover, it is 1-Lipschitz if ε is $(1-\alpha)$-Lipschitz.

If those conditions are verified, an explicit computation shows that for $\lambda \in (0, 1)$, denoting by $X_\lambda = (x_\lambda, y_\lambda)$ the unique fixed point of f_λ,

$$(x_\lambda, y_\lambda) = \left(\frac{1}{1-\lambda}\varepsilon\left(\frac{\lambda(\alpha-1)}{2(1-\alpha\lambda)}\right), \frac{\alpha-1}{2(1-\alpha\lambda)}\right).$$

By setting $g(\lambda) = \frac{\lambda(\alpha-1)}{2(1-\alpha\lambda)}$, one computes that

$$g^{-1}(\mu) = \frac{2\mu}{\alpha - 1 + 2\alpha\mu}.$$

Hence g is a bi–Lipschitz decreasing homeomorphism from $[0, 1]$ to $[-1/2, 0]$.

Now, define $h : \mathbb{R} \to \mathbb{R}$ by $h(x) = (1-x)\sin\big(\ln(|1-x|)\big)$ for $x \neq 1$ that extends by continuity with $h(1) = 0$. As

$$\forall x \neq 1, \quad h'(x) = -\sin\big(\ln(|1-x|)\big) - \cos\big(\ln(|1-x|)\big),$$

h is a Lipschitz function. It follows that for $\varepsilon_0 > 0$ small enough, the function $\varepsilon = \varepsilon_0 h \circ g^{-1}$ is $(1-\alpha)$–Lipschitz on $[-\frac{1}{2}, 0]$ and verifies $\varepsilon(-\frac{1}{2}) = 0$. Extend it by $\varepsilon(y) = \varepsilon(0)$ for $y \in [0, \frac{1}{2}]$ (Figs. 4.1 and 4.2).

For the function f associated to the latter ε, we compute that

$$\forall \lambda \in (0, 1), \quad X_\lambda = (x_\lambda, y_\lambda) = \left(\varepsilon_0 \sin\big(\ln(1-\lambda)\big), \frac{g(\lambda)}{\lambda}\right).$$

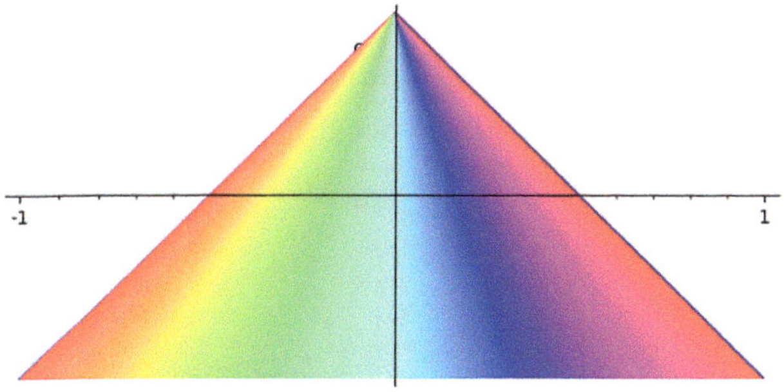

Fig. 4.1 The triangle $\mathfrak{T}$ filled with rainbow colours

Fig. 4.2 Its image by f for $\alpha = \frac{3}{4}, \varepsilon_0 = \frac{1}{10}$

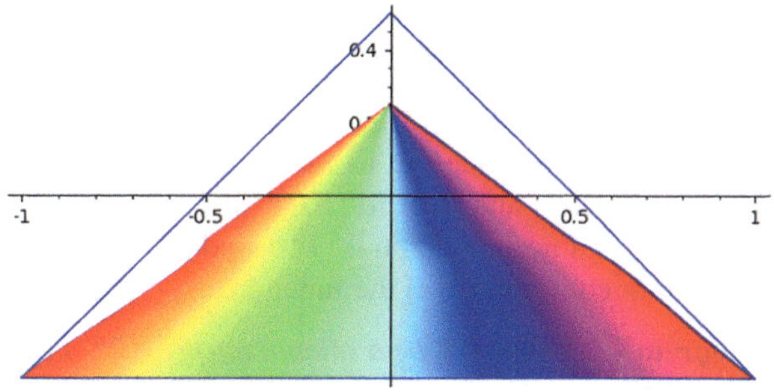

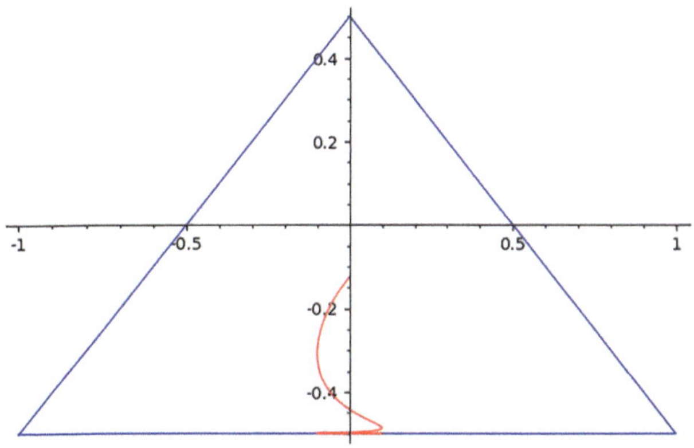

Fig. 4.3 In red, the curve of fixed points $(X_\lambda)_{\lambda \in (0,1)}$

Clearly X_λ diverges as $\lambda \to 1$. Let us also refer to [208] for other counterexamples related to the discounted equations with non–convex Hamiltonians (Figs. 4.3 and 4.4).

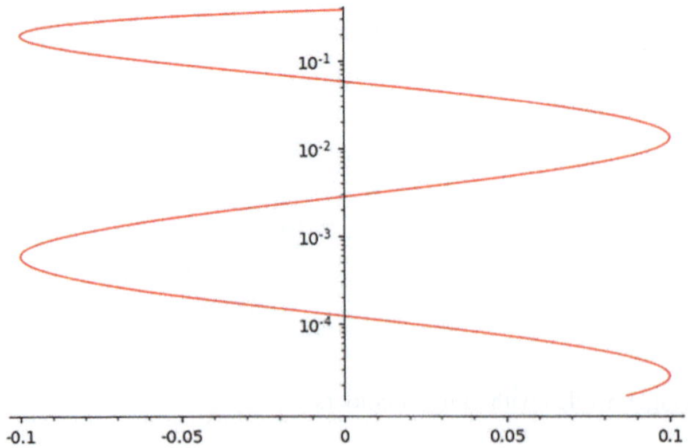

Fig. 4.4 Same curve in vertical logarithmic scale

On the positive side, let us mention another convergence result. We state it in finite dimensions and refer to [203][6] and references therein for further results. Let us recall that a norm $\|\cdot\|$ on $\mathbb{R}^n$ is called smooth if the function $x \mapsto \|x\|$ is C^1 on $\mathbb{R}^n \setminus \{0\}$ or equivalently, if the function $x \mapsto \|x\|^2$ is C^1 on $\mathbb{R}^n$.

Theorem 4.5.2 *Let $\|\cdot\|$ be a smooth norm on $\mathbb{R}^n$. Let $C \subset \mathbb{R}^n$ be a compact convex set such that $0 \in C$. Finally, let $f : C \to C$ be a 1–Lipschitz map. For all $\lambda \in (0, 1)$ we denote by $X_\lambda \in C$ the unique point such that $X_\lambda = f(\lambda X_\lambda)$. Then the family $(X_\lambda)_{\lambda\in(0,1)}$ converges as $\lambda \to 1$.*

Smoothness here is used as the unit sphere has a unique tangent linear hyperplane at each of its points. A good exercise is to prove the Theorem in the Euclidean case. If the norm comes from a scalar product $\langle\cdot,\cdot\rangle$, and if $\|x\| = 1$, this tangent hyperplane is given by the linear form $\langle x, \cdot\rangle$. In this case, it can be established that as $\lambda \to 0$, the points X_λ converge to the orthogonal projection of 0 on the set of fixed points of f.

4.6 Relations to the Classical Theory

Here again, L is a Tonelli Lagrangian on TM the tangent bundle of a smooth compact manifold M endowed with a Riemannian metric.

[6] Since writing [203], the author realized that the following result is actually a particular case of previous Theorems of Reich [184]. See also [125, 147] for many further developments.

4.6.1 Minimizing Mather Measures

The Mañé point of view: this first approach was actually introduced, after Mather's original one, by Mañé in [153, 154]. Mather then noticed that Mañé's point of view could be reformulated in a more synthetic manner and the following results were definitively written in the present form in [114].

Definition 4.6.1 A Borel probability measure μ' on TM is termed *closed* if it has finite first moment, $\int_{TM} \|v\|_x \mathrm{d}\mu'(x, v) < +\infty$ and if for all C^1 functions $f : M \to \mathbb{R}$,

$$\int_{TM} D_x f(v) \mathrm{d}\mu'(x, v) = 0.$$

We denote by $\mathcal{P}'$ the set of closed probability measures on TM.

The finite first moment condition is there so that the integral is absolutely convergent. Examples of closed measures can be constructed of the form $\nu_\gamma = \frac{1}{T}\int_0^T \delta_{(\gamma(s), \dot\gamma(s))} \mathrm{d}s$ where $\gamma : [0, T] \to M$ is a C^1 curve such that $\gamma(0) = \gamma(T)$. Indeed, if $f : M \to \mathbb{R}$ is C^1, then

$$\int_{TM} D_x f(v) \mathrm{d}\nu_\gamma(x, v) = \int_0^T D_{\gamma(s)} f\big(\dot\gamma(s)\big) \mathrm{d}s = f\big(\gamma(T)\big) - f\big(\gamma(0)\big) = 0.$$

Mañé's version of Mather measures and Mather's critical value is then:

Theorem 4.6.2 (Mañé) *The following equality holds:*

$$-\alpha(0) = \min_{\mu' \in \mathcal{P}'} \int_{TM} L(x, v) \mathrm{d}\mu'(x, v).$$

Moreover, a closed measure realizes this minimum if and only if it is supported on the Aubry set $\mathcal{A}'$. Last, a minimizing measure is automatically invariant by the Lagrangian flow φ_L.

This justifies the definition of the Mather set:

Definition 4.6.3 Let us denote by $\mathcal{P}'_0$ the set of minimizing closed probability measures, that is, closed probability measures μ' such that $\int_{TM} L(x, v) \mathrm{d}\mu' = -\alpha(0)$. The measure μ' is then said to be a *Mather measure*.

On T^*M, we define the set $\mathcal{P}^*_0 = \{\mathcal{L}_*\mu', \ \mu' \in \mathcal{P}'_0\}$.

Let us define the *Mather set* $\mathcal{M}' \subset TM$ by

$$\mathcal{M}' = \overline{\bigcup_{\mu' \in \mathcal{P}'_0} \mathrm{supp}(\mu')},$$

The *projected Mather set* is $\mathcal{M} = \pi(\mathcal{M}')$.

Finally the *Mather set in* T^*M is

$$\mathcal{M}^* = \mathcal{L}(\mathcal{M}') = \overline{\bigcup_{\mu^* \in \mathcal{P}_0^*} \operatorname{supp}(\mu^*)}.$$

It is apparent from these results that $\mathcal{M}' \subset \mathcal{A}' \subset \mathcal{L}^{-1}\big(H^{-1}(\{\alpha(0)\})\big)$ (this is Carneiro's Theorem [74]) and, as for the discrete case, and for the same reasons, $\mathcal{P}'_0$ is convex and compact and there exists one Mather measure whose support is the whole $\mathcal{M}'$. Finally, Theorem 4.1.6, stating that $\mathcal{M}$ is a uniqueness set for weak KAM solutions, holds. We do not rewrite it here.

The Mather point of view: it is more dynamical in nature, hence reminiscent of the ergodic viewpoint of Sect. 4.1.2. It also reflects Mather's original definitions as stated in [165] following his results on twist maps from [163].

Definition 4.6.4 We denote $\mathcal{P}'_L$ be the set of Borel probability measures on TM invariant by the Lagrangian flow φ_L.

The historical definition of Mather's critical constant is contained in the next result:

Proposition 4.6.5 *The critical constant is characterized by*

$$-\alpha(0) = \min_{\mu' \in \mathcal{P}'_L} \int_{TM} L(x, v)\, \mathrm{d}\mu'(x, v).$$

Moreover, minimizing measures are automatically closed, hence μ' is minimizing if and only if $\mu' \in \mathcal{P}'_0$.

From a dynamical point of view, Mather's approach is obviously more natural. However, the big drawback is that the condition of being flow invariant depends on the Lagrangian and its flow, in contrast to the condition of being closed. This is actually what motivated Mañé's change of paradigm as he wanted to study how Mather measures evolve under perturbations of a Lagrangian. It is also very useful as it applies to less regular Lagrangians and Hamiltonians.

To end this paragraph, let us pursue our systematic approach of highlighting relationships between objects coming from the classical setting and their analogues coming from the discrete setting for the time–1 action functional h_1. The main result states that projected Mather sets coincide in both settings, justifying the same notation:

Proposition 4.6.6 *Denoting by $\mathcal{M}_L$ the projected Mather set associated to L and $\mathcal{M}_{h_1}$ the projected Mather set associated to its time–1 action functional, the equality $\mathcal{M}_L = \mathcal{M}_{h_1}$ holds.*

Proof Given a point $x \in \mathcal{A}$, we will denote by $v_x \in T_x M$ the unique vector v such that $(x, v) \in \mathcal{A}'$ (Theorem 3.5.3) and by $y_x = \pi \circ \varphi_L^1(x, v_x)$ the only point such that $(x, y_x) \in \widehat{\mathcal{A}}$ (see Proposition 3.6.14). Then extend the vector–field $x \mapsto v_x$ to a Lipschitz vector–field on M for example by defining $v_x = \mathcal{L}^{-1}(D_x u)$ where u is a $C^{1,1}$ critical subsolution given by Bernard's Theorem 2.5.16 or its discrete analogue Theorem 3.4.15.

Let $\mu' \in \mathcal{P}'_0$ be a classical Mather measure. We associate to it a probability measure on $M \times M$ as follows. If $f : M \times M \to \mathbb{R}$ is a continuous function, then

$$\int_{M\times M} f(x, y)\mathrm{d}\mu(x, y) = \int_{TM} f\big(x, \pi \circ \varphi_L^1(x, v)\big)\mathrm{d}\mu'(x, v). \tag{4.3}$$

As μ' has support included in $\mathcal{A}'$, it follows that μ has support included in

$$\big\{\big(x, \pi \circ \varphi_L^1(x, v)\big), \ \ (x, v) \in \mathcal{A}'\big\} = \widehat{\mathcal{A}}.$$

Let $g : M \to \mathbb{R}$ be a continuous function, then

$$\int_{M\times M} \big(g(x) - g(y)\big)\mathrm{d}\mu(x, y) = \int_{TM} \big(g(x) - g\big(\pi \circ \varphi_L^1(x, v)\big)\big)\mathrm{d}\mu'(x, v) = 0,$$

because μ' is invariant by φ_L^1 which implies that $\int_{TM} g\big(\pi(x, v)\big)\mathrm{d}\mu' = g\big(\pi \circ \varphi_L^1(x, v)\big)\mathrm{d}\mu'(x, v)$. Hence μ is closed.

We then compute the action of μ, remembering Proposition 3.6.14:

$$\begin{aligned}
\int_{M\times M} h_1(x, y)\mathrm{d}\mu(x, y) &= \int_{TM} h_1\big(x, \pi \circ \varphi_L^1(x, v)\big)\mathrm{d}\mu'(x, v) \\
&= \int_{\mathcal{A}'} h_1\big(x, \pi \circ \varphi_L^1(x, v)\big)\, \mathrm{d}\mu'(x, v) \\
&= \int_{\mathcal{A}'} h_1\big(x, \pi \circ \varphi_L^1(x, v_x)\big)\, \mathrm{d}\pi_*\mu'(x) \\
&= \int_{\mathcal{A}'} \int_0^1 L\big(\varphi_L^s(x, v_x)\big)\, \mathrm{d}s\, \mathrm{d}\pi_*\mu'(x) \\
&= \int_0^1 \int_{\mathcal{A}'} L\big(\varphi_L^s(x, v_x)\big)\, \mathrm{d}\pi_*\mu'(x)\, \mathrm{d}s \\
&= \int_0^1 \int_{\mathcal{A}'} L(x, v_x)\mathrm{d}\pi_*\mu'(x)\, \mathrm{d}s \\
&= -\alpha(0).
\end{aligned}$$

The use of the Fubini theorem is justified by the fact that [0, 1] and $\mathcal{A}'$ are compact and L is continuous. It follows that μ is minimizing hence a discrete Mather measure. Finally, the definition of μ given by (4.3) shows that

$$\text{supp}(\mu) = \big\{\big(x, \pi \circ \varphi_L^1(x, v_x)\big),\ \ (x, v_x) \in \text{supp}(\mu')\big\}.$$

It follows that $\pi_1\big(\text{supp}(\mu)\big) = \pi\big(\text{supp}(\mu')\big)$. That being true for all measures $\mu' \in \mathcal{P}_0'$ allows to conclude that $\mathcal{M}_L \subset \mathcal{M}_{h_1}$.

Let now $\mu \in \widehat{\mathcal{P}}_0$ be a minimizing discrete Mather measure on $M \times M$. We define a measure μ_0' on TM as follows: if $f : TM \to \mathbb{R}$ is bounded and continuous,

$$\int_{TM} f(x, v) \mathrm{d}\mu_0'(x, v) = \int_{M\times M} f(x, v_x) \mathrm{d}\mu(x, y).$$

The measure μ_0' is not necessarily invariant by the whole Lagrangian flow, but it is φ_L^1–invariant. Indeed,

$$\begin{aligned}
\int_{TM} f \circ \varphi_L^1(x, v)\, \mathrm{d}\mu_0'(x, v) &= \int_{M\times M} f \circ \varphi_L^1(x, v_x)\, \mathrm{d}\mu(x, y) \\
&= \int_{\widehat{\mathcal{A}}} f \circ \varphi_L^1(x, v_x)\, \mathrm{d}\mu(x, y) \\
&= \int_{\widehat{\mathcal{A}}} f(y_x, v_{y_x})\, \mathrm{d}\mu(x, y) \\
&= \int_{\widehat{\mathcal{A}}} f(y, v_y)\, \mathrm{d}\mu(x, y) \\
&= \int_{M\times M} f(y, v_y)\, \mathrm{d}\mu(x, y) \\
&= \int_{M\times M} f(x, v_x) \mathrm{d}\mu(x, y) \\
&= \int_{TM} f(x, v)\, \mathrm{d}\mu_0'(x, v).
\end{aligned}$$

In the previous computation the second equality follows from the fact that μ is supported in $\widehat{\mathcal{A}}$, the third comes from Proposition 3.6.14, the fourth is a consequence of Proposition 3.5.3. Finally the end stems from the fact that μ is closed, applied to the function $g : x \mapsto f(x, v_x)$. It follows that the measure $\mu' = \int_0^1 (\varphi_L^s)_* \mu_0' \mathrm{d}s$ is φ_L–invariant.

Let us now prove that μ' is a classical Mather measure.

$$\begin{aligned}
\int_{TM} L(x,v)\,\mathrm{d}\mu'(x,v) &= \int_0^1 \int_{TM} L(x,v)\,\mathrm{d}(\varphi_L^s)_*\mu'_0(x,v)\,\mathrm{d}s \\
&= \int_0^1 \int_{M\times M} L\circ\varphi_L^s(x,v_x)\,\mathrm{d}\mu(x,y)\,\mathrm{d}s \\
&= \int_{M\times M} \int_0^1 L\circ\varphi_L^s(x,v_x)\,\mathrm{d}s\,\mathrm{d}\mu(x,y) \\
&= \int_{\widehat{\mathcal{A}}} \int_0^1 L\circ\varphi_L^s(x,v_x)\,\mathrm{d}s\,\mathrm{d}\mu(x,y) \\
&= \int_{\widehat{\mathcal{A}}} h_1(x,y_x)\mathrm{d}\mu(x,y) \\
&= \int_{M\times M} h_1(x,y)\,\mathrm{d}\mu(x,y) = -\alpha(0).
\end{aligned}$$

Here, it was used that μ is supported in $\widehat{\mathcal{A}}$ and that it is a discrete Mather measure. Hence μ' is minimizing as desired. Finally, it follows from its definition that

$$\mathrm{supp}(\mu') = \{\varphi_L^s(x,v_x),\quad (x,y_x)\in \mathrm{supp}(\mu),\quad s\in[0,1]\}.$$

In particular, $\pi_1\big(\mathrm{supp}(\mu)\big) \subset \pi\big(\mathrm{supp}(\mu')\big)$. As this holds for all discrete Mather measures, it comes that $\mathcal{M}_{h_1} \subset \mathcal{M}_L$.

This concludes the proof. □

Remark 4.6.7 In the previous proof, the construction associating a discrete Mather measure to a classical Mather measure, $\mu' \mapsto \mu$, is injective and the projected supports are the same.

In contrast, the reverse construction $\mu \mapsto \mu'$ may not be injective. Due to the necessity to apply the Lagrangian flow, the support increases and there is a loss of information. In other words, there may be more discrete Mather measures than classical ones; more precisely, the following inclusion holds:

$$\{\pi_*\mu',\quad \mu'\in\mathcal{P}'_0\} \subset \{\pi_{1*}\mu,\quad \mu\in\widehat{\mathcal{P}}_0\},$$

but the inclusion may be strict.

For example, as we will see later for twist maps, on a totally periodic circle which does not consist exclusively of fixed points, there is a unique classical Mather measure. In the same time, there are infinitely many discrete Mather measures obtained from averaging Dirac measures on periodic orbits.

4.6.2 The Classical Discounted Equation

Recall that for all $\ell > 0$ there exists a unique function $U_\ell : M \to \mathbb{R}$, given by Theorem 2.5.12, that solves in the viscosity sense (ℓHJ) that is, such that $\ell U_\ell(x) + H(x, D_x U_\ell) = 0$. The main result on this topic is the convergence of those functions as $\ell \to 0$ proven originally in [95] (following partial results in [143]). The result is actually obtained for Hamiltonians that are only assumed to be continuous, coercive and convex:

Theorem 4.6.8 *There exists a weak KAM solution U_0 such that $U_\ell + \frac{\alpha(0)}{\ell} \to U_0$, where the convergence takes place as $\ell \to 0$ and is uniform.*

The next lemma, called *strong comparison principle*, is an analogue to Lemma 4.2.3 and gives some more informations on Theorem 2.5.12:

Lemma 4.6.9 *Let $\ell > 0$ be a constant, let $v_1 : M \to \mathbb{R}$ (resp. $v_2 : M \to \mathbb{R}$) be a viscosity subsolution (resp. supersolution) to (ℓHJ). Then $v_1 \leqslant U_\ell \leqslant v_2$.*

The solutions, as in Lemma 4.2.6, are expressed by an explicit formula:

Lemma 4.6.10 *For any $\ell > 0$ and $x \in M$,*

$$U_\ell(x) = \min_{\substack{\gamma:(-\infty,0]\to M \\ \gamma(0)=x}} \int_{-\infty}^{0} e^{\ell s} L\big(\gamma(s), \dot\gamma(s)\big)\, \mathrm{d}s,$$

where the minimum is taken amongst absolutely continuous curves and is reached by a C^2 curve.

The selected weak KAM solution is then identified as follows

Theorem 4.6.11 *Let $\mathcal{F}' \subset \mathcal{S}'$ be the set of classical subsolutions u verifying the constraint $\int_M u(x)\mathrm{d}\pi_*\mu'(x) \leqslant 0$ for all Mather measures $\mu' \in \mathcal{P}_0'$.*

The selected weak KAM solution is then $U_0 = \sup_{u\in\mathcal{F}'} u$ where the supremum is a priori taken pointwise.

Moreover, the following alternative formula holds:

$$U_0(x) = \min_{\mu'\in\mathcal{P}_0'} \int_M h(y, x)\mathrm{d}\pi_*\mu'(y),$$

where h is again the Peierls barrier.

Let us continue with this Proposition:

Proposition 4.6.12 *There exists a Mather measure $\mu_0' \in \mathcal{P}_0'$ such that*

$$\int_X U_0(x)\mathrm{d}\pi_*\mu_0'(x) = 0.$$

Moreover, it can be imposed that μ'_0 is ergodic for the Lagrangian flow.

All those results' proofs follow closely the proofs we gave for their discrete analogue and are to be found in [95].

To conclude, as there are more measures in $\widehat{\mathcal{P}}_0$ than in $\mathcal{P}'_0$ one finds that

Proposition 4.6.13 *The following inequality holds: $u_1 \leqslant U_0$.*

However, we will provide later an example where this inequality is strict.

4.6.3 Discount for the Positive Classical L.–O. Semigroup

As is now customary, all results have a "positive" pendant by reversing time, meaning, by considering the Hamiltonian $\check{H}$. In this instance, for $\ell > 0$, define the function V_ℓ such that $-V_\ell$ is the only viscosity solution to the equation

$$\ell u(x) + \check{H}(x, D_x u) = 0, \quad x \in M. \qquad (\ell\check{\mathrm{H}}\mathrm{J})$$

Theorem 4.6.14 *There exists a positive weak KAM solution V_0 such that $V_\ell - \frac{\alpha(0)}{\ell} \to V_0$ where the convergence takes place as $\ell \to 0$ and is uniform.*

The functions V_ℓ are given by the explicit formula:

Lemma 4.6.15 *For any $\ell > 0$ and $x \in M$,*

$$V_\ell(x) = -\min_{\substack{\gamma:[0,+\infty)\to M \\ \gamma(0)=x}} \int_0^{+\infty} e^{-\ell s} L\big(\gamma(s), \dot\gamma(s)\big)\mathrm{d}s,$$

where the minimum is taken amongst absolutely continuous curves and is reached by a C^2 curve.

The limit V_0 has the following form:

Proposition 4.6.16 *Let $\mathcal{F}'^+ \subset \mathcal{S}'$ be the set of subsolutions u verifying the constraint $\int_M u(x)\mathrm{d}\pi_*\mu(x) \geqslant 0$ for all Mather measures $\mu \in \mathcal{P}'_0$.*

The limit V_0 is expressed as $V_0 = \inf\limits_{u\in\mathcal{F}'^+} u$ where the infimum is a priori taken pointwise. And finally for all $x \in M$,

$$V_0(x) = \max_{\mu\in\mathcal{P}'_0} \int_M -h(x, y)\mathrm{d}\pi_*\mu(y).$$

As for the discrete case, relations do exist between U_0 and V_0 (the proofs are similar hence omitted):

Proposition 4.6.17 *The functions U_0 and V_0 verify the inequality $U_{0|\mathcal{A}} \leqslant V_{0|\mathcal{A}}$.*

As far as conditions for U_0 and V_0 to be a conjugate pair are concerned:

Proposition 4.6.18 *The following assertions are equivalent:*

1. *The functions U_0 and V_0 form a conjugate pair,*
2. $U_{0|\mathcal{A}} = V_{0|\mathcal{A}}$,
3. $U_0 \geqslant V_0$,
4. *for all classical Mather measures $\mu' \in \mathcal{P}'_0$, the equality $\int_M U_0(x)\mathrm{d}\pi_*\mu'(x) = 0$ holds,*
5. *for all classical Mather measures $\mu' \in \mathcal{P}'_0$, the equality $\int_M V_0(x)\mathrm{d}\pi_*\mu'(x) = 0$ holds,*
6. *there exists a critical subsolution $v \in \mathcal{S}'$ such that for all Mather measures μ', the equality $\int_M v(x)\mathrm{d}\pi_*\mu'(x) = 0$ holds.*

4.6.4 Some Degenerate Discounted Hamilton–Jacobi Equations

The corresponding results are inspired by Zavidovique [205] using also methods and ideas introduced in [78] where more general problems are studied.[7] Those results hold as well for less regular Hamiltonians.

In this time–continuous setting, one still considers a continuous function $\delta : M \to \mathbb{R}$ that takes non–negative values and satisfies the condition

$$\forall \mu' \in \mathcal{P}'_0, \quad \int_M \delta(x)\, \mathrm{d}\pi_*\mu' > 0.$$

The degenerate discounted Hamilton–Jacobi equation that here studied is

$$\ell\delta(x)u(x) + H(x, D_x u) = \alpha(0), \quad x \in M. \tag{$\ell\delta$HJ}$$

To be more precise the condition prescribed in [205] is that δ is positive on the projected Aubry set. However, the more general case studied in [78] handles a wider class of perturbations of the critical equation that can be non–linear in $u(x)$. All results stated below therefore follow from those two references.

The first existence result hereafter states that our problem is well posed and is, to our knowledge original in this generality:

Theorem 4.6.19 *For all $\ell > 0$ there exists a unique viscosity solution to* ($\ell\delta$HJ) *denoted by U^δ_ℓ.*

The proof of existence uses the next lemma, which is a strong comparison principle. It is also new with such conditions on δ. It is an analogue to Theorem 4.4.4:

[7] Since then those results have been widely improved, see [97].

Lemma 4.6.20 *Let $\ell > 0$ be a constant, let $v_1 : M \to \mathbb{R}$ be a viscosity subsolution to ($\ell\delta$HJ) and $v_2 : M \to \mathbb{R}$ be a viscosity supersolution to ($\ell\delta$HJ). Then $v_1 \leqslant U_\ell \leqslant v_2$.*

The convergence result in this case is:

Theorem 4.6.21 *There exists a weak KAM solution U_0^δ such that $U_\ell^\delta \to U_0^\delta$ where the convergence takes place as $\ell \to 0$ and is uniform.*

The solutions, as in Corollary 4.4.10, are expressed by an explicit formula:

Lemma 4.6.22 *If $t > 0$ and $\gamma : [-t, 0] \to M$ is an absolutely continuous curve, we set $A_\gamma(-t) = -\int_{-t}^0 \delta \circ \gamma(s)\, \mathrm{d}s$.*

For any $\ell > 0$ and $x \in M$, if $t > 0$ then

$$U_\ell^\delta(x) = \min_{\substack{\gamma:[-t,0]\to M\\ \gamma(0)=x}} \Big\{ \exp\big(\ell A_\gamma(-t)\big) U_\ell^\delta\big(\gamma(-t)\big) + \int_{-t}^0 \exp\big(\ell A_\gamma(s)\big)\big[L\big(\gamma(s), \dot\gamma(s)\big) + \alpha(0)\big]\, \mathrm{d}s \Big\},$$

where the minimum is taken amongst absolutely continuous curves and is reached by a Lipschitz curve.

Moreover, there exists a Lipschitz curve $\gamma : (-\infty, 0] \to M$ such that $\gamma(0) = x$ and

$$U_\ell^\delta(x) = \int_{-\infty}^0 \exp\big(\ell A_\gamma(s)\big)\big[L\big(\gamma(s), \dot\gamma(s)\big) + \alpha(0)\big]\, \mathrm{d}s.$$

The selected weak KAM solution is then identified as follows:

Theorem 4.6.23 *Let $\mathcal{F}'_\delta \subset \mathcal{S}'$ be the set of classical subsolutions u verifying the constraint $\int_M \delta(x) u(x) \mathrm{d}\pi_* \mu'(x) \leqslant 0$ for all Mather measures $\mu' \in \mathcal{P}'_0$.*

The selected weak KAM solution is then $U_0^\delta = \sup_{u \in \mathcal{F}'_\delta} u$ where the supremum is taken pointwise.

Moreover, the alternative formula holds:

$$U_0(x) = \min_{\mu' \in \mathcal{P}'_0} \frac{\int_M \delta(y) h(y, x)\, \mathrm{d}\pi_*\mu'(y)}{\int_M \delta(y)\, \mathrm{d}\pi_*\mu'(y)},$$

where h is again the Peierls barrier.

Chapter 5
A Family of Examples

Abstract In this Chapter we study in detail some basic examples. Those examples all are on the 2 dimensional annulus and are of mechanical type. We characterize on those examples the weak KAM solutions selected by the various discount procedures. This shows some possible behaviors of those solutions, for instance when the cohomology varies.

We explore here explicit examples to show how the pair (u_1, v_1) may behave. Recall that on the one hand u_1 is the limit of the solutions to the discounted equations $(u_\lambda)_{\lambda\in(0,1)}$ as $\lambda \to 1$ for the negative Lax–Oleinik semigroup T^-. On the other hand, v_1 is the limit of the solutions to the discounted equations $(v_\lambda)_{\lambda\in(0,1)}$ as $\lambda \to 1$ for the positive Lax–Oleinik semigroup T^+.

As the examples presented below come from Hamiltonian systems, some familiarity with the classical theory could help the reader. The study of those examples is familiar to specialists of weak KAM theory but we have not found it written in the literature. We believe that the informations they entail is interesting and that they provide counter-examples to natural questions. At the end of the Chapter, we also address the question as to whether weak KAM solutions selected by the discounted approximation procedure in the discrete and in the continuous setting coincide.

The setting will be the one dimensional torus $\mathbb{T}^1 = \mathbb{R}/\mathbb{Z}$.

We consider a smooth potential $V : \mathbb{T}^1 \to \mathbb{R}$ that attains its maximum at exactly two points 0 and X and such that $V(0) = V(X) = 0$.

Consider the Hamiltonian function $H_0(x, p) = \frac{1}{2}p^2 + V(x)$ defined on $\mathbb{T}^1 \times \mathbb{R}$. The associated Lagrangian is then $L_0 : (x, v) \mapsto \frac{1}{2}v^2 - V(x)$. The cost function used is the time-1 action functional h_1^0 associated to H_0, as defined by (2.5). By Theorem 2.5.8, discrete and classical weak KAM solutions coincide and we will use this fact. Again, some knowledge of classical Hamilton–Jacobi equations can be useful though not necessary to read this Chapter. Moreover, as these examples

M. Zavidovique, *Discrete Weak KAM Theory*, Lecture Notes in Mathematics 2377,
https://doi.org/10.1007/978-3-031-96809-9_5

fall in the scope of Conservative Twist Maps of the annulus, the latter also illustrate results of the following and last chapter of this essay.

Let us denote by $f^{\pm} : x \mapsto \pm\sqrt{-2V(x)}$. The level set $H_0^{-1}(\{0\})$ is the union of the graphs of f^+ and f^-. Those graphs touch at $(0, 0)$ and $(X, 0)$.

A last assumption on H_0 is the following:

$$\boxed{\alpha := \frac{\int_0^X f^+(x)\mathrm{d}x}{X} < \frac{\int_X^1 f^+(x)\mathrm{d}x}{1-X} := \beta.} \tag{5.1}$$

5.1 The Study of H_0

Let $X_0^0 \in [0, X]$ and $X_1^0 \in [X, 1]$ verify that

$$\int_0^{X_0^0} f^+(x)\mathrm{d}x = \int_{X_0^0}^{X} f^+(x)\mathrm{d}x\ ; \qquad \int_X^{X_1^0} f^+(x)\mathrm{d}x = \int_{X_1^0}^{1} f^+(x)\mathrm{d}x.$$

The function u_1^0 defined by

$$u_1^0(x) = \begin{cases} \int_0^x f^+(s)\mathrm{d}s & \text{if } 0 \leqslant x \leqslant X_0^0; \\ \int_0^{X_0^0} f^+(s)\mathrm{d}s + \int_{X_0^0}^x f^-(s)\mathrm{d}s & \text{if } X_0^0 \leqslant x \leqslant X; \\ \int_X^x f^+(s)\mathrm{d}s & \text{if } X \leqslant x \leqslant X_1^0; \\ \int_X^{X_1^0} f^+(s)\mathrm{d}s + \int_{X_1^0}^x f^-(s)\mathrm{d}s & \text{if } X_1^0 \leqslant x \leqslant 1 \end{cases}$$

verifies $(u_1^0)'(s) \in H_0^{-1}(\{0\})$ at every point where the derivative exists, that is for $s \in \mathbb{T}^1 \setminus \{X_0^0, X_1^0\}$. Moreover, it is semiconcave (as seen here by the fact that at X_0^0 and X_1^0, the left derivative is bigger than the right derivative). This is enough in this context to prove that u_1^0 is a viscosity solution of the stationary Hamilton–Jacobi equation $H_0\big(x, (u_1^0)'(x)\big) = 0$ (Fig. 5.1).[1]

- It means that the critical constant is $\alpha(0) = 0$.
- The Aubry and Mather sets are included in the graph of $(u_1^0)'$ and all Hamiltonian trajectories either converge to the fixed point $(0, 0)$ or to the fixed point $(X, 0)$. It can be easily concluded from this that $\mathcal{A} = \mathcal{M} = \{0, X\}$, that classical Mather measures are convex combinations of Dirac measures $\delta_{(0,0)}$ and $\delta_{(X,0)}$ (on $T\mathbb{T}^1$), and that discrete Mather measures are convex combinations of Dirac measures, $\delta_{(0,0)}$ and $\delta_{(X,X)}$ (on $\mathbb{T}^1 \times \mathbb{T}^1$). Note that, in the present context, this illustrates

[1] Indeed, for a semiconcave function, the super solution property only has to be checked at differentiability points. For more general results see [37, 38].

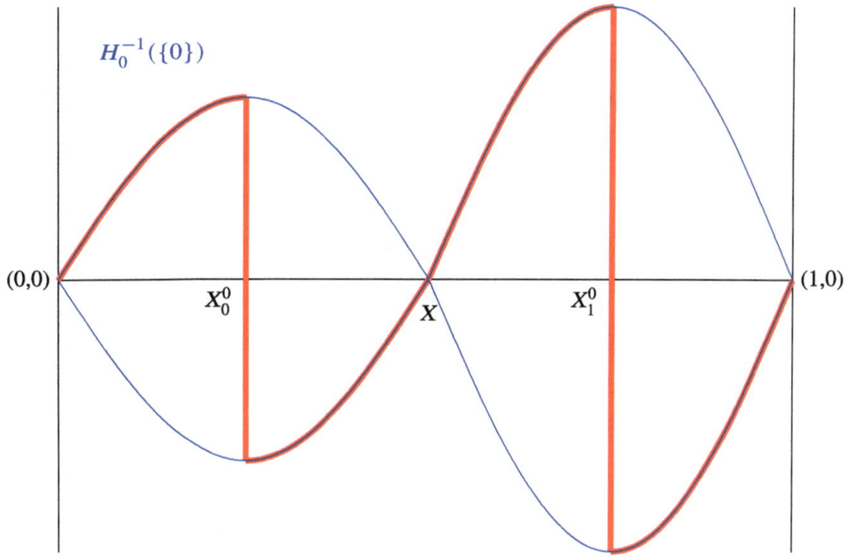

Fig. 5.1 The graph of the superdifferential $\partial^+ u_1^0$ is drawn in *red*

a classical Theorem of Carneiro for autonomous Hamiltonian systems, namely that Mather measures are supported on the critical energy level [74].

- Finally, as $u_1^0(0) = u_1^0(X) = 0$, one deduces that the function u_1^0 is indeed the weak KAM solution selected by the discounted approximation (Theorem 4.2.1). By Proposition 4.3.5, in this case, by setting v_1^0 the positive weak KAM solution selected by the discounted procedure, the pair (u_1^0, v_1^0) is a conjugate pair. Here, one easily computes that $v_1^0 = -u_1^0$.

5.2 Increasing the Cohomology Class: $c \in [0, \alpha]$

We now initiate a classical procedure in Aubry–Mather theory: changing cohomology class. This is related to the topology of the underlying space $X = \mathbb{T}^1$. This procedure is more thoroughly detailed in the final Chapter on Conservative Twist maps of the Annulus. In the Hamiltonian setting, this originates in the work of Mather [165] who noticed that suitably correcting the Lagrangian (or the Hamiltonian) by a closed 1-form does not modify the Lagrangian minimizers. The resulting objects of Aubry–Mather theory then only depend on the cohomology class of the 1-form. In the context of $\mathbb{T}^1$, the first cohomology group $H^1(\mathbb{T}^1, \mathbb{R})$ is isomorphic to $\mathbb{R}$ and if $c \in \mathbb{R}$, a representing 1-form is the constant form $x \in \mathbb{T}^1 \mapsto c$ where here c is identified to the linear form $v \in \mathbb{R} \mapsto cv$.

Let $c \in [0, \alpha]$, we consider the Hamiltonian $H_c : (x, p) \mapsto \frac{1}{2}(p+c)^2 + V(x)$. The associated Lagrangian is $L_c : (x, v) \mapsto \frac{1}{2}(v-c)^2 - \frac{1}{2}c^2 - V(x)$. The flow associated to H_c is conjugated (by a vertical translation) to that of H_0. The cost function is h_1^c, associated to the time-1 action functional of H_c. Let $X_0^c \in [0, X]$ and $X_1^c \in [X, 1]$ verify that

$$\int_0^{X_0^c} f^+(x)\mathrm{d}x - cX_0^c = \int_{X_0^c}^{X} f^+(x)\mathrm{d}x + c(X - X_0^c) \ ;$$

$$\int_X^{X_1^c} f^+(x)\mathrm{d}x - c(X_1^c - X) = \int_{X_1^c}^{1} f^+(x)\mathrm{d}x + c(1 - X_1^c).$$

The function u_1^c defined by

$$u_1^c(x) = \begin{cases} \int_0^x f^+(s)\mathrm{d}s - cx & \text{if } 0 \leqslant x \leqslant X_0^c; \\ \int_0^{X_0^c} f^+(s)\mathrm{d}s + \int_{X_0^c}^x f^-(s)\mathrm{d}s - cx & \text{if } X_0^c \leqslant x \leqslant X; \\ \int_X^x f^+(s)\mathrm{d}s - cx & \text{if } X \leqslant x \leqslant X_1^c; \\ \int_X^{X_1^c} f^+(s)\mathrm{d}s + \int_{X_1^c}^x f^-(s)\mathrm{d}s - cx & \text{if } X_1^c \leqslant x \leqslant 1; \end{cases}$$

verifies $(u_1^c)'(s) \in H_c^{-1}(\{0\})$ for $s \in \mathbb{T}^1 \setminus \{X_0^c, X_1^c\}$ and is semiconcave. As previously, this yields that u_1^c is a viscosity solution of the stationary Hamilton–Jacobi equation $H_c\big(x, (u_1^c)'(x)\big) = 0$ (Fig. 5.2).

- It means that the critical constant for the cost function h_1^c, denoted $\alpha(c)$, verifies $\alpha(c) = 0$.
- The Aubry and Mather sets are included in the graph of $(u_1^c)'$ and all Hamiltonian trajectories either converge to the fixed point $(0, -c)$ or to the fixed point $(X, -c)$. From there, it can be easily concluded that $\mathcal{A} = \mathcal{M} = \{0, X\}$ (here we drop the subscript c as the sets are independent of it), that classical Mather measures are convex combinations of Dirac measures, $\delta_{(0,0)}$ and $\delta_{(X,0)}$ (on $T\mathbb{T}^1$), and that discrete Mather measures are convex combinations of Dirac measures $\delta_{(0,0)}$ and $\delta_{(X,X)}$ (on $\mathbb{T}^1 \times \mathbb{T}^1$).
- Finally, as $u_1^c(0) = u_1^c(X) = 0$, it is deduced that the function u_1^c is indeed the weak KAM solution selected by the discounted approximation (Theorem 4.2.1).
- By Proposition 4.3.5, in this case again, by setting v_1^c the positive weak KAM solution selected by the discounted procedure, the pair (u_1^c, v_1^c) is a conjugate pair.

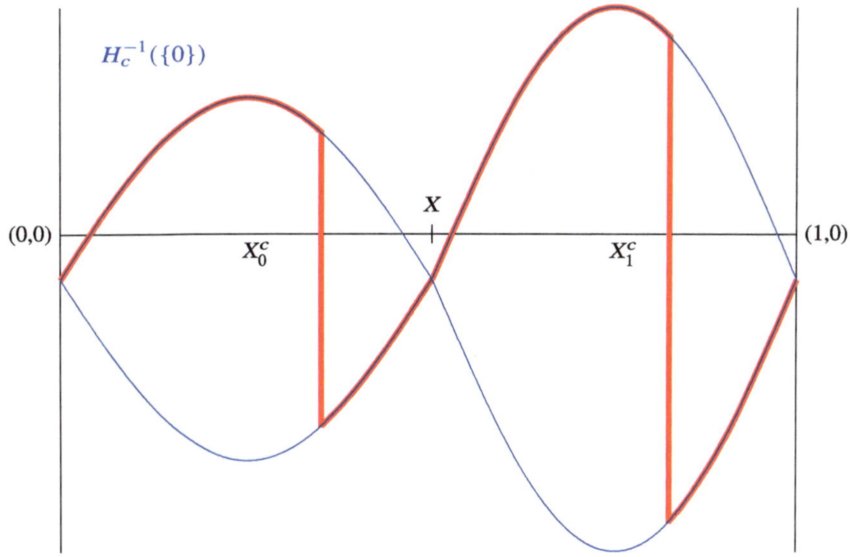

Fig. 5.2 The graph of the superdifferential $\partial^+ u_1^c$ is drawn in *red*

Here, one computes

$$v_1^c(x) = \begin{cases} \int_0^x f^-(s)\mathrm{d}s - cx & \text{if } 0 \leqslant x \leqslant \check{X}_0^c; \\ \int_0^{\check{X}_0^c} f^-(s)\mathrm{d}s + \int_{\check{X}_0^c}^x f^+(s)\mathrm{d}s - cx & \text{if } \check{X}_0^c \leqslant x \leqslant X; \\ \int_X^x f^-(s)\mathrm{d}s - cx & \text{if } X \leqslant x \leqslant \check{X}_1^c; \\ \int_X^{\check{X}_1^c} f^-(s)\mathrm{d}s + \int_{\check{X}_1^c}^x f^+(s)\mathrm{d}s - cx & \text{if } \check{X}_1^c \leqslant x \leqslant 1, \end{cases}$$

where $\check{X}_0^c \in [0, X]$ and $\check{X}_1^c \in [X, 1]$ verify

$$\int_0^{\check{X}_0^c} f^-(x)\mathrm{d}x - c\check{X}_0^c = \int_{\check{X}_0^c}^X f^-(x)\mathrm{d}x + c(X - \check{X}_0^c)\,;$$

$$\int_X^{\check{X}_1^c} f^-(x)\mathrm{d}x - c(\check{X}_1^c - X) = \int_{\check{X}_1^c}^1 f^-(x)\mathrm{d}x + c(1 - \check{X}_1^c).$$

5.3 A Change of Regime: $c \in \left(\alpha, \int_0^1 f^+(x)\mathrm{d}x\right)$

If $c \in \left(\alpha, \int_0^1 f^+(x)\mathrm{d}x\right)$. It happens that the critical constant is again 0 but it is not anymore possible to construct a critical subsolution that vanishes both at 0 and at X. Let $X^c \in [X, 1]$ verify that

$$\int_0^{X^c} f^+(x)\mathrm{d}x - cX^c = \int_{X^c}^1 f^+(x)\mathrm{d}x + c(1 - X^c).$$

The function u_1^c defined by

$$u_1^c(x) = \begin{cases} \int_0^x f^+(s)\mathrm{d}s - cx & \text{if } 0 \leqslant x \leqslant X^c; \\ \int_0^{X^c} f^+(s)\mathrm{d}s + \int_{X^c}^x f^-(s)\mathrm{d}s - cx & \text{if } X^c \leqslant x \leqslant 1. \end{cases}$$

verifies $(u_1^c)'(s) \in H_c^{-1}(\{0\})$ for $s \in \mathbb{T}^1 \setminus \{X^c\}$ and is semiconcave. As previously, this yields that u_1^c is a viscosity solution of the stationary Hamilton–Jacobi equation $H_c\big(x, (u_1^c)'(x)\big) = 0$ (Fig. 5.3).

- It means that the critical constant $\alpha(c) = 0$.
- Here again, it can be established that $\mathcal{A} = \mathcal{M} = \{0, X\}$ and that classical Mather measures are convex combinations of the Dirac measures (on $T\mathbb{T}^1$) that are $\delta_{(0,0)}$

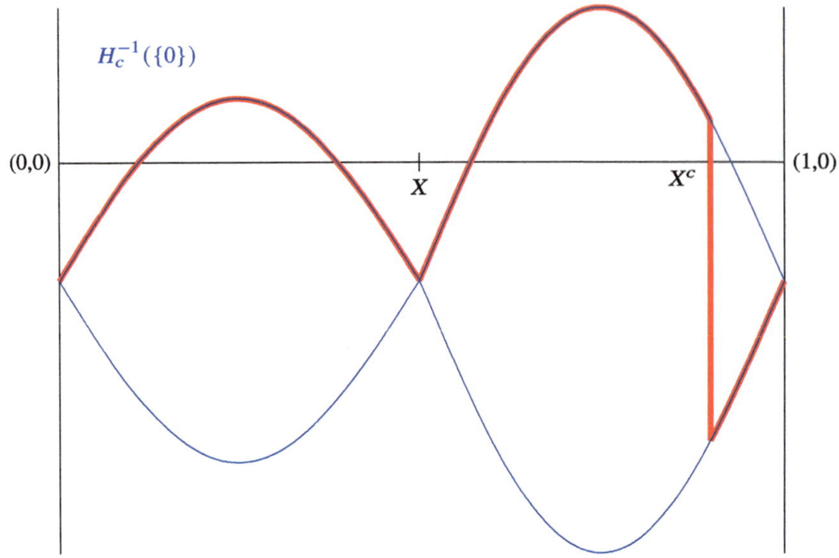

Fig. 5.3 The graph of the superdifferential $\partial^+ u_1^c$ is drawn in *red*

and $\delta_{(X,0)}$, and that discrete Mather measures are convex combinations of the Dirac measures (on $\mathbb{T}^1 \times \mathbb{T}^1$) that are $\delta_{(0,0)}$ and $\delta_{(X,X)}$. Finally, $u_1^c(0) = 0$ and $u_1^c(X) = \int_0^X f^+(s)\mathrm{d}s - cX < 0$.

- This function u_1^c is indeed the weak KAM solution selected by the discounted approximation (Theorem 4.2.1). Roughly speaking, as between 0 and X, $(u_1^c)' = f^+ - c$, it is the fastest growing weak KAM solution.
- By Proposition 4.3.5, in this case, by setting v_1^c the positive weak KAM solution selected by the discounted procedure, the pair (u_1^c, v_1^c) is NOT a conjugate pair.

Here, one computes by similar means that

$$v_1^c(x) = \begin{cases} \int_X^x f^+(s)\mathrm{d}s - c(x - X) & \text{if } 0 \leqslant x \leqslant X; \\ \int_X^x f^-(s)\mathrm{d}s - c(x - X) & \text{if } X \leqslant x \leqslant \check{X}^c; \\ \int_X^{\check{X}^c} f^-(s)\mathrm{d}s + \int_{\check{X}^c}^x f^+(s)\mathrm{d}s - c(x - X) & \text{if } \check{X}^c \leqslant x \leqslant 1; \end{cases}$$

where $\check{X}^c \in [X, 1]$ verifies

$$\int_X^0 f^+(s)\mathrm{d}s + cX = \int_X^{\check{X}^c} f^-(s)\mathrm{d}s + \int_{\check{X}^c}^1 f^+(s)\mathrm{d}s - c(1 - X).$$

Note that in this regime, for c close to $\int_0^X f^+(x)\mathrm{d}x$ the functions u_1^c and v_1^c are not ordered while, as will become clear next, for c close to $\int_0^1 f^+(x)\mathrm{d}x$ then $u_1^c < v_1^c$.

5.4 The Limiting Case: $c_0 = \int_0^1 f^+(x)\mathrm{d}x$

In this limit case, $c_0 = \int_0^1 f^+(x)\mathrm{d}x$, again the critical constant is 0 for example by invoking the continuity of Mather's α-function[2] or because we exhibit a weak KAM solution below.

Indeed the function $u_1^{c_0}$ defined by

$$\forall x \in [0, 1], \quad u_1^{c_0}(x) = \int_0^x f^+(s)\mathrm{d}s - c_0 x, \tag{5.2}$$

[2] This result is a straightforward consequence of Proposition 6.3.7 in next Chapter.

verifies $u_1^{c_0}(0) = u_1^{c_0}(1) = 0$ whence to be identified with a function on $\mathbb{T}^1$. It is C^1 (even $C^{1,1}$ in agreement with Fathi's result [108]) and a classical solution of $H_{c_0}\big(x, (u_1^{c_0})'(x)\big) = 0$, hence a weak KAM solution.

- It means that the critical constant $\alpha(c_0) = 0$. In this particular case, all viscosity subsolutions are of the form $u_1^{c_0} + K$, where $K \in \mathbb{R}$ and it can even be proved that all discrete subsolutions are of the same form.
- The situation is then different from the previous cases as the projected Aubry set is the whole torus $\mathcal{A}_{c_0} = \mathbb{T}^1$ and the classical Aubry set $\mathcal{A}^*_{c_0}$ is the whole graph of $f^+ - c_0$.
- On the contrary, as the Hamiltonian dynamics on the critical level set remains the same as in the previous examples, the invariant measures remain the same and $\mathcal{M} = \{0, X\}$.
- At last, $u_1^{c_0}(0) = 0$ and $u_1^{c_0}(X) = \int_0^X f^+(s)\mathrm{d}s - c_0 X < 0$. The function $u_1^{c_0}$ is indeed the weak KAM solution selected by the discounted approximation (Theorem 4.2.1).
- Setting $v_1^{c_0} = u_1^{c_0} - u_1^{c_0}(X)$ one obtains the weak KAM solution selected by the positive discounted approximation. Here, $u_1^{c_0} < v_1^{c_0}$ and the pair is not conjugated (in this case all negative weak KAM solutions are positive weak KAM solutions hence conjugate pairs are trivial).

5.5 Positive Rotation Numbers: $c > \int_0^1 f^+(x)\mathrm{d}x$

Let us now discuss what happens for $c > \int_0^1 f^+(x)\mathrm{d}x$. Again the cost is h_1^c associated to H_c. The behavior of weak KAM solutions and minimal trajectories are those of an area preserving twist diffeomorphism. It will be treated more thoroughly in the next Chapter but some results are briefly used here.

For $c > \int_0^1 f^+(x)\mathrm{d}x$ the situation is quite similar to the previous one. There exists a unique subsolution up to constants and therefore, up to constants, there exists a unique weak KAM solution, be it negative or positive. One such subsolution is the following: recalling that $V : \mathbb{T}^1 \to \mathbb{R}$ is the potential used in the definition of H_0, if $a > 0$, let us denote by $f_a^+ : x \mapsto \sqrt{2\big(a - V(x)\big)}$ the function whose graph is the upper part of the level set $H_0^{-1}\{a\}$. There exists a unique a_c such that $\int_0^1 f_{a_c}^+(x)\mathrm{d}x = c$. A subsolution for H_c (that is also a positive and negative weak KAM solution) is then $u : x \mapsto \int_0^x f_{a_c}^+(t)\mathrm{d}t - cx$. Therefore, the critical constant is $a_c = \alpha(c)$.

To each real number $c \in \mathbb{R}$, we associate a rotation number $\rho(c) \in \mathbb{R}$. Its projection to $\mathbb{T}^1$, written $\varrho(c)$, has the property that $\int_{\mathbb{T}^1\times\mathbb{T}^1}(y - x)\mathrm{d}\mu(x, y) = \varrho(c)$ for all Mather measure $\mu \in \widehat{\mathcal{P}}_0$. The function $\rho : \mathbb{R} \to \mathbb{R}$ is continuous and non-decreasing. It is not uniquely defined, but determined up to an integer. Here we make the choice of setting $\rho(0) = 0$. All the previous cases treated correspond to a vanishing rotation number. In our present case, the Aubry set in $T^*\mathbb{T}^1$ is the graph of the function $f^+_{\alpha(c)}$: $\mathcal{A}^*_c = \big\{\big(x, f^+_{\alpha(c)}(x) - c\big),\ \ x \in \mathbb{T}^1\big\}$ and the 2-Aubry set is (thanks to Proposition 3.6.14)

$$\widehat{\mathcal{A}}_c = \Big\{\Big(x, \pi \circ \varphi^1_{H_c}\big(x, f^+_{\alpha(c)}(x) - c\big)\Big),\quad x \in \mathbb{T}^1\Big\}.$$

Hence, the rotation number property can be rewritten

$$\begin{aligned}
&\int_{\mathbb{T}^1\times\mathbb{T}^1} (y - x)\mathrm{d}\mu(x, y)\\
&\quad = \int_{\mathbb{T}^1\times\mathbb{T}^1} \big(\pi \circ \varphi^1_{H_c}\big(x, f^+_{\alpha(c)}(x) - c\big) - x\big)\mathrm{d}\mu(x, y)\\
&\quad = \int_{\mathbb{T}^1} \big(\pi \circ \varphi^1_{H_c}\big(x, f^+_{\alpha(c)}(x) - c\big) - x\big)\mathrm{d}\pi_{1*}\mu(x, y) = \varrho(c),
\end{aligned}$$

and μ being closed yields that for all continuous function $f : \mathbb{T}^1 \to \mathbb{R}$:

$$\begin{aligned}
&\int_{\mathbb{T}^1\times\mathbb{T}^1} \big(f(y) - f(x)\big)\mathrm{d}\mu(x, y)\\
&\quad = \int_{\mathbb{T}^1\times\mathbb{T}^1} \Big(f\big(\pi \circ \varphi^1_{H_c}(x, f^+_{\alpha(c)}(x) - c)\big) - f(x)\Big)\mathrm{d}\mu(x, y)\\
&\quad = \int_{\mathbb{T}^1} \Big(f\big(\pi \circ \varphi^1_{H_c}(x, f^+_{\alpha(c)}(x) - c)\big) - f(x)\Big)\mathrm{d}\pi_{1*}\mu(x, y) = 0.
\end{aligned}$$

This means that if we set $\psi_c : x \mapsto \pi \circ \varphi^1_{H_c}\big(x, f^+_{\alpha(c)}(x) - c\big)$ (also called projected dynamics) then ψ_c is a circle diffeomorphism of rotation number $\varrho(c)$ and the measure $\pi_{1*}\mu$ is ψ_c-invariant.

We now focus on the case: $\rho(c)$ **is irrational**. In this case it is known from Poincaré-Denjoy theory that ψ_c is conjugated to a rotation of angle $\rho(c)$ (because ψ_c is C^2) and that there exists a unique ψ_c-invariant measure. Hence necessarily,

$$\pi_{1*}\mu = \frac{1}{T_c}\int_0^{T_c} \delta_{\pi\circ\varphi^s_{H_c}(0, f^+_{\alpha(c)}(0) - c)}\mathrm{d}s,$$

where T_c is the smallest positive constant such that $\varphi_{H_c}^{T_c}(0, f_{\alpha(c)}^+(0) - c) = (0, f_{\alpha(c)}^+(0) - c)$, as the latter is ψ_c-invariant.[3] Lifting things up to $\mathbb{T}^1 \times \mathbb{T}^1$, we find that

$$\mu = \frac{1}{T_c} \int_0^{T_c} \delta_{\left(\pi \circ \varphi_{H_c}^s (0, f_{\alpha(c)}^+(0)-c), \pi \circ \varphi_{H_c}^{s+1}\left(0, f_{\alpha(c)}^+(0)-c\right)\right)} \mathrm{d}s.$$

As a conclusion $\widehat{\mathcal{P}}_0 = \{\mu_c\}$ where μ_c is the previous measure and similarly, $\mathcal{P}_0^* = \{\mu_c^*\}$ where

$$\mu_c^* = \frac{1}{T_c} \int_0^{T_c} \delta_{\varphi_{H_c}^s (0, f_{\alpha(c)}^+(0)-c)} \mathrm{d}s.$$

Note that $\mathcal{P}_0^*$ is always a singleton for non 0 rotation numbers in this 1 dimensional autonomous setting. However, for rational rotation numbers, in the discrete setting, there are many Mather measures supported on the various periodic orbits.

5.5.1 Non-continuity of u_1^c with Respect to c

We now aim at studying the behavior of μ_c as $c \to c_0$:

Proposition 5.5.1 *Assume that $V''(0)V''(X) \neq 0$. Let $(c_n)_{n>0}$ be a decreasing sequence converging to c_0 such that $\rho(c_n)$ is irrational for all $n > 0$, then*

$$\mu_{c_n} \underset{n \to +\infty}{\longrightarrow} \frac{\left(\sqrt{-V''(0)}\right)^{-1}}{\left(\sqrt{-V''(0)}\right)^{-1} + \left(\sqrt{-V''(X)}\right)^{-1}} \delta_{(0,0)} + \frac{\left(\sqrt{-V''(X)}\right)^{-1}}{\left(\sqrt{-V''(0)}\right)^{-1} + \left(\sqrt{-V''(X)}\right)^{-1}} \delta_{(X,X)}.$$

Proof As the set of probability measures on $X \times X$ is compact, to prove the result one just needs to prove that any converging subsequence of $(\mu_{c_n})_{n>0}$ has the announced limit. Hence without loss of generality, let us assume that $(\mu_{c_n})_{n>0}$ is a converging sequence. We now prove it converges to $\frac{\left(\sqrt{-V''(0)}\right)^{-1}}{\left(\sqrt{-V''(0)}\right)^{-1} + \left(\sqrt{-V''(X)}\right)^{-1}} \delta_{(0,0)} + \frac{\left(\sqrt{-V''(X)}\right)^{-1}}{\left(\sqrt{-V''(0)}\right)^{-1} + \left(\sqrt{-V''(X)}\right)^{-1}} \delta_{(X,X)}$.

[3] It can be proven that $\rho(c) = T_c^{-1}$.

We know from [165] for example, or Proposition 6.3.7 below, that the function $c \mapsto \alpha(c)$ is convex and continuous. Hence the limit of the sequence $(\mu_{c_n})_{n>0}$ is a Mather measure, therefore of the form $\beta_0\delta_{(0,0)} + \beta_X\delta_{(X,X)}$ with $0 \leqslant \beta_0, \beta_X \leqslant 1$ and $\beta_0 + \beta_X = 1$.

With this information at hand, we will in fact only study the measures $\pi_{1*}\mu_{c_n}$ and prove they converge to $\frac{\left(\sqrt{-V''(0)}\right)^{-1}}{\left(\sqrt{-V''(0)}\right)^{-1}+\left(\sqrt{-V''(X)}\right)^{-1}}\delta_0+\frac{\left(\sqrt{-V''(X)}\right)^{-1}}{\left(\sqrt{-V''(0)}\right)^{-1}+\left(\sqrt{-V''(X)}\right)^{-1}}\delta_X$.

Let $\gamma_n : t \mapsto \pi \circ \varphi^t_{H_{c_n}}(0, f^+_{\alpha(c_n)}(0) - c_n)$. By looking at the Hamiltonian equations (2.4) and recalling that H_{c_n} is constant on a Hamiltonian trajectory, one finds that

$$\forall t \in \mathbb{R}, \quad \dot{\gamma}_n(t) = \sqrt{2\big(\alpha(c_n) - V\big(\gamma_n(t)\big)\big)}. \tag{5.3}$$

The coefficients β_0 and β_X are proportional to the relative amount of time that the trajectory γ_n stays respectively in a neighborhood of 0 and X, as $n \to +\infty$.

Until the end of this proof, let us no longer think of points on the circle $\mathbb{T}^1$ but by lifting to $\mathbb{R}$, still keeping the same notations. Hence the function V is now a 1-periodic function on $\mathbb{R}$. Integrating (5.3), one computes that if $x < y$, the time it takes γ_n to go from x to y is

$$t_{x,y} = \int_x^y \frac{\mathrm{d}s}{\sqrt{2\big(\alpha(c_n) - V(s)\big)}}. \tag{5.4}$$

In particular, $T_{c_n} = \int_0^1 \frac{\mathrm{d}s}{\sqrt{2\left(\alpha(c_n)-V(s)\right)}}$.

Let $0 < \varepsilon < \max\{-V''(0), -V''(X)\}$. Let us consider $\eta > 0$ such that,

$$[|x| < \eta] \Longrightarrow \left[\left|V(x) - V''(0)\frac{x^2}{2}\right| < \varepsilon\frac{x^2}{2}\right],$$

and

$$[|x - X| < \eta] \Longrightarrow \left[\left|V(x) - V''(X)\frac{(x-X)^2}{2}\right| < \varepsilon\frac{(x-X)^2}{2}\right].$$

We now split the integral defining T_{c_n} into the 4 following pieces:

$$\begin{aligned} T_{c_n} &= \int_0^1 \frac{\mathrm{d}s}{\sqrt{2\left(\alpha(c_n) - V(s)\right)}} \\ &= \underbrace{\int_{-\eta}^{\eta} \frac{\mathrm{d}s}{\sqrt{2\left(\alpha(c_n) - V(s)\right)}}}_{①} + \underbrace{\int_{X-\eta}^{X+\eta} \frac{\mathrm{d}s}{\sqrt{2\left(\alpha(c_n) - V(s)\right)}}}_{②} \end{aligned}$$

$$+\underbrace{\int_{\eta}^{X-\eta}\frac{\mathrm{d}s}{\sqrt{2\left(\alpha(c_n)-V(s)\right)}}}_{③}+\underbrace{\int_{X+\eta}^{1-\eta}\frac{\mathrm{d}s}{\sqrt{2\left(\alpha(c_n)-V(s)\right)}}}_{④}. \tag{5.5}$$

Let $M > 0$ be a constant independent of n such that $|③| + |④| < M$ for all $n > 0$. Such an M exists as the denominators appearing in the integrals are uniformly positive.

Let us now study and estimate ①. From the definition of η the following inequalities are inferred:

$$\underbrace{\int_{-\eta}^{\eta}\frac{\mathrm{d}s}{\sqrt{2\left(\alpha(c_n)+\left(\varepsilon-V''(0)\right)\frac{s^2}{2}\right)}}}_{⑤}$$

$$\leqslant \int_{-\eta}^{\eta}\frac{\mathrm{d}s}{\sqrt{2\left(\alpha(c_n)-V(s)\right)}}$$

$$\leqslant \underbrace{\int_{-\eta}^{\eta}\frac{\mathrm{d}s}{\sqrt{2\left(\alpha(c_n)+\left(-V''(0)-\varepsilon\right)\frac{s^2}{2}\right)}}}_{⑥}. \tag{5.6}$$

Terms ⑤ and ⑥ can be integrated explicitly, let us deal with ⑤.

$$\int_{-\eta}^{\eta}\frac{\mathrm{d}s}{\sqrt{2\left(\alpha(c_n)+\left(\varepsilon-V''(0)\right)\frac{s^2}{2}\right)}}=\frac{\sqrt{\frac{2\alpha(c_n)}{\varepsilon-V''(0)}}}{\sqrt{2\alpha(c_n)}}\int_{-\eta\sqrt{\frac{\varepsilon-V''(0)}{2\alpha(c_n)}}}^{\eta\sqrt{\frac{\varepsilon-V''(0)}{2\alpha(c_n)}}}\frac{\mathrm{d}t}{\sqrt{1+t^2}}$$

$$=\frac{1}{\sqrt{\varepsilon-V''(0)}}\left[\ln\left(\sqrt{1+t^2}+t\right)\right]_{-\eta\sqrt{\frac{\varepsilon-V''(0)}{2\alpha(c_n)}}}^{\eta\sqrt{\frac{\varepsilon-V''(0)}{2\alpha(c_n)}}}$$

$$=\frac{1}{\sqrt{\varepsilon-V''(0)}}\ln\left[\frac{\sqrt{\frac{\left(\varepsilon-V''(0)\right)\eta^2}{2\alpha(c_n)}+1}+\eta\sqrt{\frac{\varepsilon-V''(0)}{2\alpha(c_n)}}}{\sqrt{\frac{\left(\varepsilon-V''(0)\right)\eta^2}{2\alpha(c_n)}+1}-\eta\sqrt{\frac{\varepsilon-V''(0)}{2\alpha(c_n)}}}\right]$$

$$=\frac{1}{\sqrt{\varepsilon-V''(0)}}\ln\left[\left(\sqrt{\frac{\left(\varepsilon-V''(0)\right)\eta^2}{2\alpha(c_n)}+1}+\eta\sqrt{\frac{\varepsilon-V''(0)}{2\alpha(c_n)}}\right)^2\right]$$

$$
\begin{aligned}
&= \frac{1}{\sqrt{\varepsilon - V''(0)}} \Big\{ - \ln(2) - \ln\big(\alpha(c_n)\big) \\
&\quad + \ln\left[\sqrt{\big(\varepsilon - V''(0)\big)\eta^2 + 2\alpha(c_n)} + \eta\sqrt{\varepsilon - V''(0)}\right]\Big\} \\
&\underset{n\to+\infty}{\sim} \frac{-\ln\big(\alpha(c_n)\big)}{\sqrt{\varepsilon - V''(0)}}.
\end{aligned}
$$

The last relation uses the continuity of α and the consecutive limit: $\lim_{n\to+\infty} \alpha(c_n) = 0$.

The same computation for ⑥ yields

$$
\int_{-\eta}^{\eta} \frac{\mathrm{d}s}{\sqrt{2\left(\alpha(c_n) + \big(-V''(0) - \varepsilon\big)\frac{s^2}{2}\right)}} \underset{n\to+\infty}{\sim} \frac{-\ln\big(\alpha(c_n)\big)}{\sqrt{-\varepsilon - V''(0)}}.
$$

As for ② the same strategy is adopted:

$$
\begin{aligned}
&\underbrace{\int_{X-\eta}^{X+\eta} \frac{\mathrm{d}s}{\sqrt{2\left(\alpha(c_n) + \big(\varepsilon - V''(X)\big)\frac{s^2}{2}\right)}}}_{⑦} \\
&\quad \leqslant \int_{X-\eta}^{X+\eta} \frac{\mathrm{d}s}{\sqrt{2\left(\alpha(c_n) - V(s)\right)}} \\
&\quad \leqslant \underbrace{\int_{X-\eta}^{X+\eta} \frac{\mathrm{d}s}{\sqrt{2\left(\alpha(c_n) + \big(-V''(X) - \varepsilon\big)\frac{s^2}{2}\right)}}}_{⑧}.
\end{aligned}
\tag{5.7}
$$

Similar computations yield

$$
\int_{X-\eta}^{X+\eta} \frac{\mathrm{d}s}{\sqrt{2\left(\alpha(c_n) + \big(\varepsilon - V''(X)\big)\frac{s^2}{2}\right)}} \underset{n\to+\infty}{\sim} \frac{-\ln\big(\alpha(c_n)\big)}{\sqrt{\varepsilon - V''(X)}},
$$

and

$$\int_{X-\eta}^{X+\eta} \frac{\mathrm{d}s}{\sqrt{2\left(\alpha(c_n) + \big(-V''(X) - \varepsilon\big)\frac{s^2}{2}\right)}} \underset{n\to+\infty}{\sim} \frac{-\ln\big(\alpha(c_n)\big)}{\sqrt{-\varepsilon - V''(X)}}.$$

Let now $f : \mathbb{R} \to \mathbb{R}$ be a continuous 1-periodic function that is constant on $[-\eta, \eta]$ and on $[X - \eta, X + \eta]$, with $0 \leqslant \min\big(f(0), f(X)\big)$. We know that

$$\int_{[0,1]} f(s)\mathrm{d}\pi_{1*}\mu_{c_n}(s) \underset{n\to+\infty}{\longrightarrow} \beta_0 f(0) + \beta_X f(X).$$

Gathering the previous computations we infer that

$$\begin{aligned}\frac{f(0)① + f(X)② - \|f\|_\infty M}{T_{c_n}} &\leqslant \int_{[0,1]} f(s)\mathrm{d}\pi_{1*}\mu_{c_n}(s)\\ &\leqslant \frac{f(0)① + f(X)② + \|f\|_\infty M}{T_{c_n}}.\end{aligned}$$

And letting $n \to +\infty$ one discovers that

$$\begin{aligned}&\frac{\dfrac{f(0)}{\sqrt{\varepsilon - V''(0)}} + \dfrac{f(X)}{\sqrt{\varepsilon - V''(X)}}}{\dfrac{1}{\sqrt{-\varepsilon - V''(0)}} + \dfrac{1}{\sqrt{-\varepsilon - V''(X)}}}\\ &\leqslant \beta_0 f(0) + \beta_X f(X)\\ &\leqslant \frac{\dfrac{f(0)}{\sqrt{-\varepsilon - V''(0)}} + \dfrac{f(X)}{\sqrt{-\varepsilon - V''(X)}}}{\dfrac{1}{\sqrt{\varepsilon - V''(0)}} + \dfrac{1}{\sqrt{-\varepsilon - V''(X)}}}.\end{aligned}$$

This being true for all $\varepsilon > 0$ and all non-negative $f(0)$ and $f(X)$, the lemma is proved. □

At last, we can deduce the following:

Proposition 5.5.2 *Assume that $H_0 : (x, p) \mapsto \frac{1}{2}p^2 + V(x)$ is a Hamiltonian such that $V : \mathbb{T}^1 \to \mathbb{R}$ is smooth, non-positive, and verifies $V^{-1}\{0\} = \{0, X\}$ for some $X \in \mathbb{T}^1 \setminus \{0\}$. Assume moreover that $V''(0) \neq V''(X)$ are both negative and again*

$$\alpha := \frac{\int_0^X f^+(x)\mathrm{d}x}{X} < \frac{\int_X^1 f^+(x)\mathrm{d}x}{1 - X} := \beta.$$

Then, using the previous notations, if $(c_n)_{n>0}$ is a decreasing sequence converging to c_0 such that $\rho(c_n)$ is irrational for all $n > 0$, the family of functions $(u_1^{c_n})_{n>0}$ does not converge to $u_1^{c_0}$.

Proof Using that a limit of weak KAM solutions is a weak KAM solution and the previous Proposition 5.5.1, one proves that $(u_1^{c_n})_{n>0}$ converges to the unique weak KAM solution u, for H_{c_0} such that

$$\frac{\left(\sqrt{-V''(0)}\right)^{-1}}{\left(\sqrt{-V''(0)}\right)^{-1}+\left(\sqrt{-V''(X)}\right)^{-1}}u(0)+\frac{\left(\sqrt{-V''(X)}\right)^{-1}}{\left(\sqrt{-V''(0)}\right)^{-1}+\left(\sqrt{-V''(X)}\right)^{-1}}u(X)=0.$$

Recall that at cohomology c_0, there is a unique weak KAM solution, up to constants. Moreover, the condition $\alpha < \beta$ ensures that it is not possible to have $u(0) = u(X) = 0$. Hence $u(0)u(X) < 0$. On the contrary, as $u_1^{c_0}$ is given by the formula (5.2):

$$\forall x \in [0, 1], \quad u_1^{c_0}(x) = \int_0^x f^+(s)\mathrm{d}s - c_0 x,$$

we have $u_1^{c_0}(0)u_1^{c_0}(X) = 0$.

□

5.5.2 A Situation Where $u_1^c \neq U_0^c$

We come back to Proposition 4.6.13. More precisely, we answer by the negative the natural question: does the discounted procedure select the same weak KAM solution in the discrete setting and in the continuous setting?

We now focus our attention on the unique real number $c_{\frac{1}{2}} \in \mathbb{R}$ such that $\rho(c_{\frac{1}{2}}) = \frac{1}{2}$. Note that $c_{\frac{1}{2}} > 0$ and that $\mathcal{A}^*_{c_{\frac{1}{2}}} \subset H^{-1}(\{\alpha(c_{\frac{1}{2}})\})$. It is actually the upper connected component of this level set of H. Moreover, it can be characterized as follows:

Proposition 5.5.3 *Let $\Phi_{H_{c_{\frac{1}{2}}}} : \mathbb{R}^2 \to \mathbb{R}^2$ denote the lift of $\varphi_{H_{c_{\frac{1}{2}}}}$ that fixes the point $(0, -c_{\frac{1}{2}})$ and $P : \mathbb{R}^2 \to \mathbb{T}^1 \times \mathbb{R}$ the canonical projection. Then*

$$\mathcal{A}^*_{c_{\frac{1}{2}}} = P\Big(\big\{(x, p) \in \mathbb{R}^2, \quad \Phi^2_{H_{c_{\frac{1}{2}}}}(x, p) = (x + 1, p)\big\}\Big).$$

Sketch of Proof There are many possible ways to tackle this Proposition according to the property of the system used. The proof is essentially given in [6, Proposition 15] and very much related to [4, Proposition 2] that proves a version of the result, in arbitrary dimension, by using C^0-integrability in a neighborhood of $\mathcal{A}^*_{c_{\frac{1}{2}}}$. It can also be deduced from the Implicit function Theorem, proving that the right hand side of the above equality is a manifold.

Let us sketch a proof using a stronger integrability, reminiscent of the Arnol'd–Liouville Theorem [100]. Let us call B the right hand side of the equality to prove. A first step is that B is a (potentially partial) graph over $\mathbb{T}^1$. We distinguish three cases remembering that Hamiltonian orbits are included in level sets of H.

- if $H(x,p) < \max V = 0$ then the orbit $\{\varphi^s_{H_{c_{\frac{1}{2}}}}(x,p), \quad s \in \mathbb{R}\}$ projects on an interval strictly included in $\mathbb{T}^1$ hence $(x,p) \notin B$.
- if $H(x,p) \geqslant \max V$ with $p \leqslant -c_{\frac{1}{2}}$. Let $(\tilde{x},p) \in \mathbb{R}^2$ such that $P(\tilde{x},p) = (x,p)$. Then if we define for $s \in \mathbb{R}$, $\Phi^s_{H_{c_{\frac{1}{2}}}}(\tilde{x},p) = (\tilde{x}_s, p_s)$, one computes that $s \mapsto \tilde{x}_s$ is non-increasing, hence $(x,p) \notin B$.
- if $H(x,p) \geqslant \max V$ with $p \geqslant -c_{\frac{1}{2}}$, meaning that $p \geqslant \sqrt{-2V(x)} - c_{\frac{1}{2}}$. We fix here $\tilde{x} \in \mathbb{R}$ and let p vary. Looking at the Hamiltonian equations and more precisely computing the time t_p such that $\Phi^{t_p}_{H_{c_{\frac{1}{2}}}}(\tilde{x},p) = (\tilde{x}+1, p+1)$ with Eq. (5.4), it can be seen that $p \mapsto t_p$ is decreasing. Hence there is at most one p such that $t_p = 2$.

Now recall that $\mathcal{A}^*_{c_{\frac{1}{2}}} = \big\{\big(x, f^+_{\alpha(c_{\frac{1}{2}})}(x) - c_{\frac{1}{2}}\big), \quad x \in \mathbb{T}^1\big\}$. We denote by $\mathfrak{A}$ the lift of $\mathcal{A}^*_{c_{\frac{1}{2}}}$ to $\mathbb{R}^2$, that is invariant under $\Phi_{H_{c_{\frac{1}{2}}}}$. We define the map $g : \mathbb{R} \to \mathbb{R}$ by

$$\forall x \in \mathbb{R}, \quad \Phi^1_{H_{c_{\frac{1}{2}}}}\big(x, f^+_{\alpha(c_{\frac{1}{2}})}(x) - c_{\frac{1}{2}}\big) = \big(g(x), f^+_{\alpha(c_{\frac{1}{2}})} \circ g(x) - c_{\frac{1}{2}}\big).$$

The function g is the lift of a circle diffeomorphism and its rotation number here is $\rho(c_{\frac{1}{2}}) = \frac{1}{2}$. It follows from Poincaré's theory of rotation numbers that there exists a real number $x_0 \in \mathbb{R}$ such that $g^2(x_0) = x_0 + 1$. Let now $x \in \mathbb{R}$ be any real number. There is a time $t \in \mathbb{R}$ such that $\big(x, f^+_{\alpha(c_{\frac{1}{2}})}(x) - c_{\frac{1}{2}}\big) = \Phi^t_{H_{c_{\frac{1}{2}}}}\big(x_0, f^+_{\alpha(c_{\frac{1}{2}})}(x_0) - c_{\frac{1}{2}}\big)$. It follows that

$$\begin{aligned}
\Phi^2_{H_{c_{\frac{1}{2}}}}\big(x, f^+_{\alpha(c_{\frac{1}{2}})}(x) - c_{\frac{1}{2}}\big) &= \Phi^2 \circ \Phi^t_{H_{c_{\frac{1}{2}}}}\big(x_0, f^+_{\alpha(c_{\frac{1}{2}})}(x_0) - c_{\frac{1}{2}}\big)\\
&= \Phi^t \circ \Phi^2_{H_{c_{\frac{1}{2}}}}\big(x_0, f^+_{\alpha(c_{\frac{1}{2}})}(x_0) - c_{\frac{1}{2}}\big)\\
&= \Phi^t \circ \Phi^2_{H_{c_{\frac{1}{2}}}}\big(x_0+1, f^+_{\alpha(c_{\frac{1}{2}})}(x_0) - c_{\frac{1}{2}}\big)\\
&= \big(x+1, f^+_{\alpha(c_{\frac{1}{2}})}(x) - c_{\frac{1}{2}}\big).
\end{aligned}$$

Thus it has been proven that $\mathfrak{A} \subset B$ and that the right hand side is a partial graph while the left hand side is a full graph. Hence both terms are equal. □

Let us define the map $h : \mathbb{T}^1 \to \mathbb{R}$ by

$$\forall x \in \mathbb{T}^1, \quad \varphi^1_{H_{c_{\frac{1}{2}}}}\big(x, f^+_{\alpha(c_{\frac{1}{2}})}(x) - c_{\frac{1}{2}}\big) = \big(h(x), f^+_{\alpha(c_{\frac{1}{2}})} \circ h(x) - c_{\frac{1}{2}}\big).$$

The function h has $g : \mathbb{R} \to \mathbb{R}$ as a lift. It follows from the previous result and Theorem 4.1.3 that for all $x \in \mathbb{T}^1$, the measure $\mu_x = \frac{1}{2}\Big(\delta_{\big(x,h(x)\big)} + \delta_{\big(h(x),x\big)}\Big)$ is a discrete Mather measure.

Lemma 5.5.4 *Assume that* $u_1^{c_{\frac{1}{2}}} = U_0^{c_{\frac{1}{2}}}$, *then for all* $x \in \mathbb{T}^1$, $u_1^{c_{\frac{1}{2}}}(x) = -u_1^{c_{\frac{1}{2}}} \circ g(x)$.

Proof From the definitions of $u_1^{c_{\frac{1}{2}}}$ and $U_0^{c_{\frac{1}{2}}}$, and from the previous discussions result that

$$\forall x \in \mathbb{T}^1, \quad \int_{\mathbb{T}^1} u_1^{c_{\frac{1}{2}}}(s)\mathrm{d}\pi_{1*}\mu_x(s) = \frac{1}{2}\Big(u_1^{c_{\frac{1}{2}}}(x) + u_1^{c_{\frac{1}{2}}} \circ g(x)\Big) \leqslant 0. \tag{5.8}$$

Moreover, using Proposition 4.6.12 and the fact that there is a single classical Mather measure, we obtain that

$$\int_{\mathbb{T}^1} u_1^{c_{\frac{1}{2}}}(x)\mathrm{d}\pi_*\mu^*_{c_{\frac{1}{2}}}(x) = 0.$$

Recalling the definition of $\mu^*_{c_{\frac{1}{2}}}$ we find:

$$\begin{aligned}
0 &= \int_{\mathbb{T}^1} u_1^{c_{\frac{1}{2}}}(x)\mathrm{d}\pi_*\mu^*_{c_{\frac{1}{2}}}(x) = \int_0^2 u_1^{c_{\frac{1}{2}}}\Big(\pi \circ \varphi^s_{H_{c_{\frac{1}{2}}}}\big(0, f^+_{\alpha(c_{\frac{1}{2}})}(0) - c_{\frac{1}{2}}\big)\Big)\mathrm{d}s \\
&= \int_0^1 \Big[u_1^{c_{\frac{1}{2}}}\Big(\pi \circ \varphi^s_{H_{c_{\frac{1}{2}}}}\big(0, f^+_{\alpha(c_{\frac{1}{2}})}(0) - c_{\frac{1}{2}}\big)\Big) \\
&\quad + u_1^{c_{\frac{1}{2}}}\Big(\pi \circ \varphi^{s+1}_{H_{c_{\frac{1}{2}}}}\big(0, f^+_{\alpha(c_{\frac{1}{2}})}(0) - c_{\frac{1}{2}}\big)\Big)\Big]\mathrm{d}s \\
&= \int_0^1 2\int_{\mathbb{T}^1} u_1^{c_{\frac{1}{2}}}(x)\mathrm{d}\pi_{1*}\mu_{\pi\circ\varphi^s_{H_{c_{\frac{1}{2}}}}\big(0, f^+_{\alpha(c_{\frac{1}{2}})}(0)-c_{\frac{1}{2}}\big)}(x)\mathrm{d}s \leqslant 0.
\end{aligned}$$

It follows that all inequalities in (5.8) are equalities, hence the result.

□

It is deduced that under the hypotheses of the previous Lemma, $u_1^{c_{\frac{1}{2}}}(x)$ and $u_1^{c_{\frac{1}{2}}} \circ g(x)$ must have opposite signs for all x. An example in which it is not the case is provided in [21, Appendix A.2.]. We give below a different simple situation where this cannot happen:

Proposition 5.5.5 *Let $V : \mathbb{T}^1 \to \mathbb{R}$ be a non-constant $\frac{1}{2}$-periodic function. Then for the associated Hamiltonian $H_{c_{\frac{1}{2}}}$, it holds $u_1^{c_{\frac{1}{2}}} \neq U_0^{c_{\frac{1}{2}}}$.*

Proof Indeed, in this case, both functions $u_1^{c_{\frac{1}{2}}}$ and $U_0^{c_{\frac{1}{2}}}$ are also $\frac{1}{2}$-periodic (this follows from the uniqueness of the solutions to the discounted equations that hence must be $\frac{1}{2}$-periodic). Consequently if they coincide, the previous lemma tells us that they must be identically 0. This is clearly not the case. □

5.6 Concluding Example

Due to the simple structure of the Aubry set and of the set of minimizing measures in all previous examples, one can check that u_1^c and v_1^c form a conjugate pair "up to a constant". More precisely, the modified pair $\big(u_1^c - u_1^c(0), v_1^c - v_1^c(0)\big)$ is a conjugate pair in all the previous examples. One may wonder if this is always the case.

As a matter of fact, the answer is again negative. We propose here a slightly (but not too much) sophisticated example shedding light on this fact. The computations are not carried on fully and left to the reader. We hope the previous examples give enough insight to make what follows quite straightforward.

Once again we consider that $H_0 : (x, p) \mapsto \frac{1}{2}p^2 + V(x)$ is a Hamiltonian such that $V : \mathbb{T}^1 \to \mathbb{R}$ is smooth, non-positive, and verifies $V^{-1}\{0\} = \{0, X_1, X_2\}$ for some $0 < X_1 < X_2 < 1 \in \mathbb{T}^1$. Assume again that

$$\underbrace{\frac{\int_0^{X_1} f^+(x)\mathrm{d}x}{X_1}}_{=\alpha} < \underbrace{\frac{\int_{X_1}^{X_2} f^+(x)\mathrm{d}x}{X_2 - X_1}}_{=\beta} < \underbrace{\frac{\int_{X_2}^{1} f^+(x)\mathrm{d}x}{1 - X_2}}_{=\gamma}.$$

If $c \in \mathbb{R}$ we again denote by $H_c : (x, p) \mapsto \frac{1}{2}(p + c)^2 + V(x)$ and respectively by u_1^c and v_1^c the negative and positive weak KAM solutions selected by the discounted procedure for the time-1 minimal action functional associated to H_c. Reasoning as in the previous sections, one checks that for $\alpha < c < \min(\beta, c_0)$, the function u_1^c verifies $u_1^c(0) = u_1^c(X_2) = 0$ and $u_1^c(X_1) < 0$ while the function v_1^c verifies $v_1^c(X_1) = v_1^c(X_2) = 0$ and $v_1^c(0) > 0$. Hence u_1^c and v_1^c are not conjugated "up to a constant".

Chapter 6
Twist Maps

Abstract This final Chapter applies and develop discrete weak KAM theory in the context conservative twist maps of the annulus. This is the original setting in which Aubry–Mather theory was conceived and strangely, weak KAM solutions were not very much studied in this setting until quite recently. In this Chapter we present the very special features of Aubry and Mather sets in this low dimensional setting. We then study and characterize weak KAM solutions for those twist maps, we describe the shape of their derivative. Finally (partially without proof) we present results related to, more or less weak, integrability properties of twist maps.

The results of discrete weak KAM theory will now be applied to the particular and founding case of Exact Conservative Twist Maps of the annulus. Excellent surveys on the subject are [16, 30, 172] and we refer to those references for classical results that we leave without proof. The aim of this section is to present some of the results of [21–23] obtained in collaboration with Marie–Claude Arnaud, shedding light on the structure of weak KAM solutions and minimizing orbits for exact conservative twist maps. Before giving the precise definition of an exact conservative twist map, let us emphasize that the examples of the previous section are closely related to such transformations. Indeed, for a Tonelli Hamiltonian on $T^*\mathbb{T}^1$, the Hamiltonian flow φ_H^s is an exact conservative twist map for small times $s > 0$. Hence the time 1 map φ_H^1 is a composition of a finite number of conservative twist maps.

6.1 Definitions and Variational Structure

In the rest of this section $\mathbb{T}^1 = \mathbb{R}/\mathbb{Z}$ is the circle, the 2–dimensional annulus is denoted by $\mathbb{A} = T^*\mathbb{T}^1 = \mathbb{T}^1 \times \mathbb{R}$. The points of that annulus are denoted by $(\theta, r) \in \mathbb{A}$. Throughout this section we will often consider objects coming from $\mathbb{A}$ lifted to $\mathbb{R}^2$, its universal cover, or from $\mathbb{T}^1$, lifted to $\mathbb{R}$. When done so, a $\sim$ will be added to the original name. For example if $g : \mathbb{T}^1 \to \mathbb{R}$ is any function then $\tilde{g} : \mathbb{R} \to \mathbb{R}$ is the lift of g.

M. Zavidovique, *Discrete Weak KAM Theory*, Lecture Notes in Mathematics 2377,
https://doi.org/10.1007/978-3-031-96809-9_6

When dealing with products, $\mathbb{T}^1 \times \mathbb{T}^1$, $\mathbb{T}^1 \times \mathbb{R}$, or $\mathbb{R} \times \mathbb{R}$, the notations π_1 and π_2 stand for the projections on the first and second variable.

6.1.1 Definition and Birkhoff's Theorem

Definition 6.1.1 An exact conservative twist map of the annulus (abbreviated ECTM) is a C^1–diffeomorphism $f : \mathbb{A} \to \mathbb{A}$ such that

1. f is isotopic to the identity map.
2. f is exact symplectic: by denoting $f(\theta, r) = \big(\Theta(\theta, r), R(\theta, r)\big)$ the 1–form $R\mathrm{d}\Theta - r\mathrm{d}\theta$ is exact.
3. f twists verticals to the right: if $\tilde{f} = (\widetilde{\Theta}, R) : \mathbb{R}^2 \to \mathbb{R}^2$ is a lift of f to the universal cover of $\mathbb{A}$ then for all $\tilde{\theta} \in \mathbb{R}$, the map $r \mapsto \widetilde{\Theta}(\tilde{\theta}, r)$ is an increasing C^1–diffeomorphism of $\mathbb{R}$.

Remark 6.1.2

1. The first property means that there is a continuous path of diffeomorphisms of the annulus $(f_s)_{s\in[0,1]}$ such that f_0 is the identity map and $f_1 = f$. In other topological words, f preserves both ends of the annulus in the sense that uniformly in θ, $\lim_{r\to+\infty} R(\theta, r) = +\infty$ and $\lim_{r\to-\infty} R(\theta, r) = -\infty$.
2. The second point means that there is a function $S : \mathbb{A} \to \mathbb{R}$, called *generating function*, such that

$$\mathrm{d}S = R\mathrm{d}\Theta - r\mathrm{d}\theta = R\Big(\frac{\partial\Theta}{\partial\theta}\mathrm{d}\theta + \frac{\partial\Theta}{\partial r}\mathrm{d}r\Big) - r\mathrm{d}\theta. \tag{6.1}$$

The generating function is defined up to a constant.

Another way of formulating this is to say that if we denote by $\lambda = r\mathrm{d}\theta$ (the 1–form called Liouville form) then $f^*\lambda - \lambda$ is exact:

This has two major implications.

(i) It is proved by using Stokes' formula that if C is a C^1 essential circle, meaning an injective C^1 closed curve going around the cylinder, then the algebraic area between C and $f(C)$ is 0. This means that points are not globally shifted up or down by f in the annulus and that it is worth looking for invariant compact sets.

(ii) The second is that f preserves the canonical symplectic 2–form which is here the Lebesgue area form. Indeed, as this symplectic form is $\mathrm{d}\lambda = \mathrm{d}r \wedge \mathrm{d}\theta$, one finds that

$$0 = \mathrm{dd}S = \mathrm{d}R \wedge \mathrm{d}\Theta - \mathrm{d}r \wedge \mathrm{d}\theta = f^*(\mathrm{d}r \wedge \mathrm{d}\theta) - \mathrm{d}r \wedge \mathrm{d}\theta.$$

3. The last point implies that for $\tilde{\theta}_0 \in \mathbb{R}$ and $\widetilde{\Theta}_0 \in \mathbb{R}$ there is a unique $r \in \mathbb{R}$ such that $\widetilde{\Theta}(\tilde{\theta}_0, r) = \widetilde{\Theta}_0$. It means that $(\tilde{\theta}, r) \mapsto \big(\tilde{\theta}, \widetilde{\Theta}(\tilde{\theta}, r)\big)$ is a C^1–diffeomorphism that plays the role of Legendre transform and $(\tilde{\theta}, \widetilde{\Theta})$ will serve as coordinates of $\mathbb{R}^2$. The latter powerful idea is behind the whole variational structure of twist maps.

   The twist condition implies that

$$\forall(\theta_0, r_0) \in \mathbb{A}, \quad \frac{\partial \Theta}{\partial r}(\theta_0, r_0) > 0. \tag{6.2}$$

   In fact, it is sometimes replaced by the stronger uniform condition that there exists $\varepsilon > 0$ such that $\frac{\partial \Theta}{\partial r}(\theta_0, r_0) > \varepsilon$ for all $(\theta_0, r_0) \in \mathbb{A}$.

One can at this stage state an important dynamical property of ECTM on invariant curves (see [59]):

Theorem 6.1.3 (Birkhoff) *Let $C \subset \mathbb{A}$ be a continuous essential circle invariant by f, i.e. a continuous embedding of $\mathbb{T}^1$ that is not homotopic to a point and such that $f(C) = C$. Then there exists a Lipschitz function $g : \mathbb{T}^1 \to \mathbb{R}$ such that $C = \{(\theta, g(\theta)),\ \theta \in \mathbb{T}^1\}$ is the graph of g.*

This important Theorem uses all properties of the ECTM. Indeed in [150, Proposition 5.13] are examples of non conservative twist maps leaving invariant essential circles that are not graphs. It was generalized a few decades later to higher dimensional settings by Arnaud and Arnaud, Venturelli [8, 20].

6.1.2 The Generating Function, Properties and Consequences

Until the end of this section, we choose once and for all a generating function $S : \mathbb{A} \to \mathbb{R}$ (see Remark 6.1.2), $\widetilde{S} : \mathbb{R}^2 \to \mathbb{R}$ is the lift of S and let us choose $\tilde{f} = (\widetilde{\Theta}, R) : \mathbb{R}^2 \to \mathbb{R}^2$ a lift of f. By the twist condition (Definition 6.1.1 point 3), the map $\mathcal{L} : (\tilde{\theta}, r) \mapsto \big(\tilde{\theta}, \widetilde{\Theta}(\tilde{\theta}, r)\big)$ is a C^1–diffeomorphism of $\mathbb{R}^2$ (see the links with Definition 3.5.2). In the sequel, $(\tilde{\theta}, \widetilde{\Theta})$ are systematically used as coordinates, meaning that the function $\widetilde{S} \circ \mathcal{L}^{-1}$ is considered instead of $\widetilde{S}$. For readability issues, it is still written $\widetilde{S}(\tilde{\theta}, \widetilde{\Theta})$.

The properties of f translate into the following features of S:

Proposition 6.1.4 *The function S, through $\widetilde{S}$, verifies the following:*

1. *The function $\widetilde{S}$ is C^2 and periodic:*

$$\forall(\tilde{\theta}, \widetilde{\Theta}) \in \mathbb{R}^2, \quad \widetilde{S}(\tilde{\theta} + 1, \widetilde{\Theta} + 1) = \widetilde{S}(\tilde{\theta}, \widetilde{\Theta}).$$

2. *For all* $(\tilde{\theta}, \widetilde{\Theta}, r, R) \in \mathbb{R}^4$,

$$\tilde{f}(\tilde{\theta}, r) = (\widetilde{\Theta}, R) \iff \begin{cases} r = -\dfrac{\partial \widetilde{S}}{\partial \tilde{\theta}}(\tilde{\theta}, \widetilde{\Theta}), \\ R = \dfrac{\partial \widetilde{S}}{\partial \widetilde{\Theta}}(\tilde{\theta}, \widetilde{\Theta}). \end{cases} \tag{6.3}$$

3. *The twist condition translates as follows:*
   - *for* $\tilde{\theta} \in \mathbb{R}$ *fixed, the map* $\widetilde{\Theta} \mapsto \frac{\partial \widetilde{S}}{\partial \tilde{\theta}}(\tilde{\theta}, \widetilde{\Theta})$ *is a decreasing* C^1*–diffeomorphism of* $\mathbb{R}$*;*
   - *for* $\widetilde{\Theta} \in \mathbb{R}$ *fixed, the map* $\tilde{\theta} \mapsto \frac{\partial \widetilde{S}}{\partial \widetilde{\Theta}}(\tilde{\theta}, \widetilde{\Theta})$ *is a decreasing* C^1*–diffeomorphism of* $\mathbb{R}$*.*

Let us comment on the previous proposition.

Remark 6.1.5

1. It can actually be established that given a function $\widetilde{S}$ satisfying the three points of the previous proposition the associated function $\tilde{f}$ is a lift of an ECTM.
2. The second point of Proposition 6.1.4 can be directly read on Eq. (6.1).
3. The two items of point 3 are equivalent. The first one results from the fact that the map $\Theta \mapsto -\frac{\partial \widetilde{S}}{\partial \tilde{\theta}}(\tilde{\theta}, \widetilde{\Theta})$ is the inverse of the map $r \mapsto \widetilde{\Theta}(\tilde{\theta}, r)$.

   The second one is an emanation of the fact that if f is an ECTM that twists verticals to the right then f^{-1} is an ECTM that twists verticals to the left. A direct consequence of any of these two facts is that

$$\forall (\tilde{\theta}_0, \widetilde{\Theta}_0) \in \mathbb{R}^2, \qquad \frac{\partial^2 \widetilde{S}}{\partial \tilde{\theta} \partial \widetilde{\Theta}}(\tilde{\theta}_0, \widetilde{\Theta}_0) < 0. \tag{6.4}$$

   Another important aftermath is that $\widetilde{S}$ is superlinear, meaning that

$$\lim_{|\tilde{\theta} - \widetilde{\Theta}| \to +\infty} \frac{\widetilde{S}(\tilde{\theta}, \widetilde{\Theta})}{|\tilde{\theta} - \widetilde{\Theta}|} = +\infty. \tag{6.5}$$

It is proven in [162] (see also [30, 163]) that finite compositions of ECTM also possess a similar generating function. Hence all variational results relying solely on the generating function and its minimizers also apply to such finite compositions of ECTM.

Let us now define the notion of minimizing chains and sequences.

Definition 6.1.6

1. Let $a < b$ be two integers such that $b - a > 1$ and $(\tilde{\theta}_i)_{a \leqslant i \leqslant b} \in \mathbb{R}^{b-a+1}$. The chain $(\tilde{\theta}_i)_{a \leqslant i \leqslant b}$ is termed *minimizing* if

$$\sum_{i=a}^{b-1} \widetilde{S}(\tilde{\theta}_i, \tilde{\theta}_{i+1}) = \min_{\substack{\{x_i \in \mathbb{R},\ a \leqslant i \leqslant b\} \\ x_a = \tilde{\theta}_a, x_b = \tilde{\theta}_b}} \sum_{i=a}^{b-1} \widetilde{S}(x_i, x_{i+1}).$$

2. An infinite sequence $(\tilde{\theta}_i)_{-\infty < i \leqslant b}$ or $(\tilde{\theta}_i)_{a \leqslant i < +\infty}$ or $(\tilde{\theta}_i)_{i \in \mathbb{Z}}$ is minimizing if all its finite subchains are minimizing.

Note that if $(\tilde{\theta}_i)_{a \leqslant i \leqslant b} \in \mathbb{R}^{b-a+1}$ is minimizing between $\tilde{\theta}_a$ and $\tilde{\theta}_b$ and if $a \leqslant a' < b' - 1 < b' \leqslant b$ then the subchain $(\tilde{\theta}_i)_{a' \leqslant i \leqslant b'} \in \mathbb{R}^{b'-a'+1}$ is minimizing between $\tilde{\theta}_{a'}$ and $\tilde{\theta}_{b'}$ which justifies the second definition.

The first order necessary condition for a chain to be minimizing translates as follows: let $(\tilde{\theta}_i)_{i \in I}$ be a minimizing chain or sequence and $i_0 \in I$ not be an extremity of I, then

$$\frac{\partial \widetilde{S}}{\partial \widetilde{\Theta}}(\tilde{\theta}_{i_0-1}, \tilde{\theta}_{i_0}) + \frac{\partial \widetilde{S}}{\partial \tilde{\theta}}(\tilde{\theta}_{i_0}, \tilde{\theta}_{i_0+1}) = 0. \tag{6.6}$$

This, together with Eq. (6.3), yields the following proposition:

Proposition 6.1.7 *Let $(\tilde{\theta}_i)_{i \in I}$ be a minimizing chain or sequence and for all $i \in I$, let*

$$r_i = -\frac{\partial \widetilde{S}}{\partial \tilde{\theta}}(\tilde{\theta}_i, \tilde{\theta}_{i+1}) = \frac{\partial \widetilde{S}}{\partial \widetilde{\Theta}}(\tilde{\theta}_{i-1}, \tilde{\theta}_i),$$

(where only the well defined term is taken if i is an extremity of I). Then $(\tilde{\theta}_i, r_i)_{i \in I}$ is a piece of orbit of $\tilde{f}$ in the sense that for all $i \in I$ such that $i + 1 \in I$, then $\tilde{f}(\tilde{\theta}_i, r_i) = (\tilde{\theta}_{i+1}, r_{i+1})$.

In particular, a minimizing chain is uniquely determined by two consecutive terms $(\tilde{\theta}_i, \tilde{\theta}_{i+1})$.

Let us then introduce the notion of crossing and its consequences:

Definition 6.1.8 Two chains $(\tilde{\theta}_i)_{i \in I}$ and $(\tilde{\theta}'_i)_{i \in I}$ are said to cross

- at some index $i_0 \in I$ if $\tilde{\theta}_{i_0} = \tilde{\theta}'_{i_0}$;
- between two indices $i_0 \in I$ and $i_0 + 1 \in I$ if $(\tilde{\theta}_{i_0} - \tilde{\theta}'_{i_0})(\tilde{\theta}_{i_0+1} - \tilde{\theta}'_{i_0+1}) < 0$.

Remark 6.1.9 If $(\tilde{\theta}_i)_{i \in I}$ and $(\tilde{\theta}'_i)_{i \in I}$ are distinct minimizing chains that cross at some index $i_0 \in I$ that is an interior index, then the twist condition implies that

$$(\tilde{\theta}_{i_0-1} - \tilde{\theta}'_{i_0-1})(\tilde{\theta}_{i_0+1} - \tilde{\theta}'_{i_0+1}) < 0. \tag{6.7}$$

Indeed, let $r_{i_0} = -\frac{\partial \widetilde{S}}{\partial \tilde{\theta}}(\tilde{\theta}_{i_0}, \tilde{\theta}_{i_0+1})$ and $r'_{i_0} = -\frac{\partial \widetilde{S}}{\partial \tilde{\theta}}(\tilde{\theta}'_{i_0}, \tilde{\theta}'_{i_0+1})$. Necessarily $r_{i_0} \neq r'_{i_0}$ because otherwise $(\tilde{\theta}_i)_{i\in I}$ and $(\tilde{\theta}'_i)_{i\in I}$ are the projections of the same orbit of $\tilde{f}$ (Proposition 6.1.7). Assume then for example that $r_{i_0} > r'_{i_0}$. We deduce from the twist condition that $\tilde{\theta}_{i_0+1} > \tilde{\theta}'_{i_0+1}$ and that $\tilde{\theta}_{i_0-1} < \tilde{\theta}'_{i_0-1}$.

Inequality (6.7) is often taken as the definition of crossing at i_0.

One can then state Aubry & Le Daeron's Fundamental Lemma:

Lemma 6.1.10 *Let* $(\tilde{\theta}, \tilde{\theta}', \widetilde{\Theta}, \widetilde{\Theta}') \in \mathbb{R}^4$ *such that* $(\tilde{\theta} - \tilde{\theta}')(\widetilde{\Theta} - \widetilde{\Theta}') < 0$. *Then*

$$\widetilde{S}(\tilde{\theta}, \widetilde{\Theta}) + \widetilde{S}(\tilde{\theta}', \widetilde{\Theta}') > \widetilde{S}(\tilde{\theta}, \widetilde{\Theta}') + \widetilde{S}(\tilde{\theta}', \widetilde{\Theta}).$$

Proof It follows from the chain of equalities (in which we use the notations $\widetilde{\Theta}_t = t\widetilde{\Theta}' + (1-t)\widetilde{\Theta}$ and $\tilde{\theta}_t = t\tilde{\theta}' + (1-t)\tilde{\theta}$),

$$\begin{aligned}
&\widetilde{S}(\tilde{\theta}, \widetilde{\Theta}') - \widetilde{S}(\tilde{\theta}, \widetilde{\Theta}) + \widetilde{S}(\tilde{\theta}', \widetilde{\Theta}) - \widetilde{S}(\tilde{\theta}', \widetilde{\Theta}') \\
&= \int_0^1 \Big[\frac{\partial \widetilde{S}}{\partial \widetilde{\Theta}}(\tilde{\theta}, \widetilde{\Theta}_t) - \frac{\partial \widetilde{S}}{\partial \widetilde{\Theta}}(\tilde{\theta}', \widetilde{\Theta}_t)\Big]\mathrm{d}t \cdot (\widetilde{\Theta}' - \widetilde{\Theta}) \\
&= \int_0^1 \int_0^1 \frac{\partial^2 \widetilde{S}}{\partial \tilde{\theta} \partial \widetilde{\Theta}}(\tilde{\theta}_s, \widetilde{\Theta}_t)\mathrm{d}s\mathrm{d}t \cdot (\tilde{\theta} - \tilde{\theta}')(\widetilde{\Theta}' - \widetilde{\Theta}) < 0.
\end{aligned}$$

The last inequality is a consequence of the crossing hypothesis and of (6.4). □

We may now state Aubry's Non–crossing Lemma:

Proposition 6.1.11 *Let* $(\tilde{\theta}_i)_{i\in I}$ *and* $(\tilde{\theta}'_i)_{i\in I}$ *be two distinct minimizing chains. Then one of the following holds*

- $(\tilde{\theta}_i)_{i\in I}$ *and* $(\tilde{\theta}'_i)_{i\in I}$ *don't cross,*
- $(\tilde{\theta}_i)_{i\in I}$ *and* $(\tilde{\theta}'_i)_{i\in I}$ *cross exactly once,*
- $I = [a, b]$ *is a finite interval and* $(\tilde{\theta}_i)_{i\in I}$ *and* $(\tilde{\theta}'_i)_{i\in I}$ *cross exactly twice, at* a *and* b.

In the last case, both $(\tilde{\theta}_i)_{i\in I}$ *and* $(\tilde{\theta}'_i)_{i\in I}$ *are maximal in the sense that neither one is a strict subchain of a minimizing chain.*

Proof To prove this proposition, let us assume that $(\tilde{\theta}_i)_{i\in I}$ and $(\tilde{\theta}'_i)_{i\in I}$ cross (at least) twice and that one of those crossing is not an extremity of I. There are several cases to deal with, we only cover two of them and let the other ones as an exercise as the ideas are the same.

First case: there are $\alpha < \beta$ such that $(\tilde{\theta}_i)_{i\in I}$ and $(\tilde{\theta}'_i)_{i\in I}$ cross between α and $\alpha+1$ and between β and $\beta + 1$. Then define two chains $(\hat{\theta}_i)_{i\in[\alpha,\beta+1]}$ and $(\hat{\theta}'_i)_{i\in[\alpha,\beta+1]}$ as follows:

$$\hat{\theta}_i = \begin{cases} \tilde{\theta}'_i & \text{if } i \in [\alpha+1, \beta] \\ \tilde{\theta}_i & \text{if } i \in \{\alpha, \beta+1\}; \end{cases}$$

$$\hat{\theta}'_i = \begin{cases} \tilde{\theta}_i & \text{if } i \in [\alpha+1, \beta] \\ \tilde{\theta}'_i & \text{if } i \in \{\alpha, \beta+1\}. \end{cases}$$

Note that $(\tilde{\theta}_i)_{i\in[\alpha,\beta+1]}$ and $(\hat{\theta}_i)_{i\in[\alpha,\beta+1]}$ have same endpoints and so do $(\tilde{\theta}'_i)_{i\in[\alpha,\beta+1]}$ and $(\hat{\theta}'_i)_{i\in[\alpha,\beta+1]}$ hence

$$\sum_{i=\alpha}^{\beta} \widetilde{S}(\tilde{\theta}_i, \tilde{\theta}_{i+1}) \leqslant \sum_{i=\alpha}^{\beta} \widetilde{S}(\hat{\theta}_i, \hat{\theta}_{i+1}); \quad \text{and} \quad \sum_{i=\alpha}^{\beta} \widetilde{S}(\tilde{\theta}'_i, \tilde{\theta}'_{i+1}) \leqslant \sum_{i=\alpha}^{\beta} \widetilde{S}(\hat{\theta}'_i, \hat{\theta}'_{i+1}).$$

Moreover

$$\begin{aligned} &\sum_{i=\alpha}^{\beta} \widetilde{S}(\tilde{\theta}_i, \tilde{\theta}_{i+1}) + \sum_{i=\alpha}^{\beta} \widetilde{S}(\tilde{\theta}'_i, \tilde{\theta}'_{i+1}) - \sum_{i=\alpha}^{\beta} \widetilde{S}(\hat{\theta}_i, \hat{\theta}_{i+1}) - \sum_{i=\alpha}^{\beta} \widetilde{S}(\hat{\theta}'_i, \hat{\theta}'_{i+1}) \\ &\quad = \widetilde{S}(\tilde{\theta}_\alpha, \tilde{\theta}_{\alpha+1}) + \widetilde{S}(\tilde{\theta}_\beta, \tilde{\theta}_{\beta+1}) + \widetilde{S}(\tilde{\theta}'_\alpha, \tilde{\theta}'_{\alpha+1}) + \widetilde{S}(\tilde{\theta}'_\beta, \tilde{\theta}'_{\beta+1}) \\ &\qquad - \widetilde{S}(\tilde{\theta}_\alpha, \tilde{\theta}'_{\alpha+1}) - \widetilde{S}(\tilde{\theta}_\beta, \tilde{\theta}'_{\beta+1}) - \widetilde{S}(\tilde{\theta}'_\alpha, \tilde{\theta}_{\alpha+1}) - \widetilde{S}(\tilde{\theta}'_\beta, \tilde{\theta}_{\beta+1}) > 0; \end{aligned} \tag{6.8}$$

where the last inequality is obtained by two applications of Aubry's Fundamental Lemma 6.1.10. This contradicts either the fact that $(\tilde{\theta}_i)_{i\in[\alpha,\beta+1]}$ is minimizing or that $(\tilde{\theta}'_i)_{i\in[\alpha,\beta+1]}$ is minimizing.

Second case: there are $\alpha < \beta$ such that $(\tilde{\theta}_i)_{i\in I}$ and $(\tilde{\theta}'_i)_{i\in I}$ cross at α and at β and such that $\alpha - 1 \in I$.

Then define two chains $(\hat{\theta}_i)_{i\in[\alpha-1,\beta]}$ and $(\hat{\theta}'_i)_{i\in[\alpha-1,\beta]}$ as follows:

$$\hat{\theta}_i = \begin{cases} \tilde{\theta}'_i & \text{if } i \in [\alpha, \beta] \\ \tilde{\theta}_i & \text{if } i \in \{\alpha-1, \alpha\}; \end{cases}$$

$$\hat{\theta}'_i = \begin{cases} \tilde{\theta}_i & \text{if } i \in [\alpha, \beta] \\ \tilde{\theta}'_i & \text{if } i \in \{\alpha-1, \alpha\}. \end{cases}$$

We purposely insist on the fact that $\hat{\theta}_\alpha = \hat{\theta}'_\alpha = \tilde{\theta}_\alpha = \tilde{\theta}'_\alpha$. Note that $(\tilde{\theta}_i)_{i\in[\alpha-1,\beta]}$ and $(\hat{\theta}_i)_{i\in[\alpha-1,\beta]}$ have same endpoints and so do $(\tilde{\theta}'_i)_{i\in[\alpha-1,\beta]}$ and $(\hat{\theta}'_i)_{i\in[\alpha-1,\beta]}$. Hence

$$\sum_{i=\alpha-1}^{\beta-1} \widetilde{S}(\tilde{\theta}_i, \tilde{\theta}_{i+1}) \leqslant \sum_{i=\alpha-1}^{\beta-1} \widetilde{S}(\hat{\theta}_i, \hat{\theta}_{i+1}); \text{ and } \sum_{i=\alpha-1}^{\beta-1} \widetilde{S}(\tilde{\theta}'_i, \tilde{\theta}'_{i+1}) \leqslant \sum_{i=\alpha-1}^{\beta-1} \widetilde{S}(\hat{\theta}'_i, \hat{\theta}'_{i+1}).$$

Moreover the definitions of $(\hat{\theta}_i)_{i\in[\alpha-1,\beta]}$ and $(\hat{\theta}'_i)_{i\in[\alpha-1,\beta]}$ yield

$$\sum_{i=\alpha}^{\beta}\widetilde{S}(\tilde{\theta}_i,\tilde{\theta}_{i+1})+\sum_{i=\alpha}^{\beta}\widetilde{S}(\tilde{\theta}'_i,\tilde{\theta}'_{i+1})-\sum_{i=\alpha}^{\beta}\widetilde{S}(\hat{\theta}_i,\hat{\theta}_{i+1})-\sum_{i=\alpha}^{\beta}\widetilde{S}(\hat{\theta}'_i,\hat{\theta}'_{i+1})=0.$$

It follows that both inequalities above are equalities and that $(\tilde{\theta}_i)_{i\in[\alpha-1,\beta]}$ and $(\hat{\theta}_i)_{i\in[\alpha-1,\beta]}$ are both minimizing. As they coincide for both indices $\alpha-1$ and α, they are equal (see Proposition 6.1.7). The same argument shows that $(\tilde{\theta}'_i)_{i\in[\alpha-1,\beta]}$ and $(\hat{\theta}'_i)_{i\in[\alpha-1,\beta]}$ are equal, and finally we have proved that $(\tilde{\theta}_i)_{i\in[\alpha-1,\beta]}$ and $(\tilde{\theta}'_i)_{i\in[\alpha-1,\beta]}$ are equal. This in turn implies that $(\tilde{\theta}_i)_{i\in I}=(\tilde{\theta}'_i)_{i\in I}$ which is a contradiction.

The last assertion of the Proposition remains to be proven. The previous argument could be adapted here. Let us though propose another one. Assume by contradiction that $(\tilde{\theta}_i)_{i\in[a,b]}$ and $(\tilde{\theta}'_i)_{i\in[a-1,b]}$ are minimizing and verify $\tilde{\theta}_a=\tilde{\theta}'_a$, $\tilde{\theta}_b=\tilde{\theta}'_b$ (the other cases are treated the same way). Then by the minimization hypothesis it is inferred that

$$\sum_{i=a}^{b-1}\widetilde{S}(\tilde{\theta}_i,\tilde{\theta}_{i+1})=\sum_{i=a}^{b-1}\widetilde{S}(\tilde{\theta}'_i,\tilde{\theta}'_{i+1}).$$

Moreover, note that the chain $(\tilde{\theta}'_{a-1},\tilde{\theta}_a,\tilde{\theta}_{a+1})$ is not minimizing, otherwise there would be equality $\tilde{\theta}_{a+1}=\tilde{\theta}'_{a+1}$ contradicting the beginning of the Proposition. It follows there exists $\hat{\theta}\in\mathbb{R}$ such that

$$\widetilde{S}(\tilde{\theta}'_{a-1},\hat{\theta})+\widetilde{S}(\hat{\theta},\tilde{\theta}_{a+1})<\widetilde{S}(\tilde{\theta}'_{a-1},\tilde{\theta}_a)+\widetilde{S}(\tilde{\theta}_a,\tilde{\theta}_{a+1}).$$

The chain $(\hat{\theta}_i)_{i\in[a-1,b]}$ is then defined as follows:

$$\hat{\theta}_i=\begin{cases}\tilde{\theta}'_{a-1} \text{ if } i=a-1\\ \hat{\theta} \text{ if } i=a\\ \tilde{\theta}_i \text{ if } i\in[a+1,b];\end{cases}$$

inducing that

$$\sum_{i=a-1}^{b-1}\widetilde{S}(\hat{\theta}_i,\hat{\theta}_{i+1})=\widetilde{S}(\tilde{\theta}'_{a-1},\hat{\theta})+\widetilde{S}(\hat{\theta},\tilde{\theta}_{a+1})+\sum_{i=a+1}^{b-1}\widetilde{S}(\tilde{\theta}_i,\tilde{\theta}_{i+1})$$

$$<\widetilde{S}(\tilde{\theta}'_{a-1},\tilde{\theta}_a)+\widetilde{S}(\tilde{\theta}_a,\tilde{\theta}_{a+1})+\sum_{i=a+1}^{b-1}\widetilde{S}(\tilde{\theta}_i,\tilde{\theta}_{i+1})=\sum_{i=a-1}^{b-1}\widetilde{S}(\tilde{\theta}'_i,\tilde{\theta}'_{i+1}).$$

This contradicts the fact that $(\tilde{\theta}'_i)_{i\in[a-1,b]}$ is minimizing. □

The previous non–crossing Proposition can be enforced.

Definition 6.1.12 When I is suitably infinite, one says that two chains $(\tilde{\theta}_i)_{i\in I}$ and $(\tilde{\theta}'_i)_{i\in I}$ are respectively

- α–asymptotic if $\lim_{i\to-\infty} |\tilde{\theta}_i - \tilde{\theta}'_i| = 0$; we then say $(\tilde{\theta}_i)_{i\in I}$ and $(\tilde{\theta}'_i)_{i\in I}$ cross at $-\infty$;
- ω–asymptotic if $\lim_{i\to+\infty} |\tilde{\theta}_i - \tilde{\theta}'_i| = 0$; we then say $(\tilde{\theta}_i)_{i\in I}$ and $(\tilde{\theta}'_i)_{i\in I}$ cross at $+\infty$.

Using this terminology, we state without proof (see for instance [30, Lemma 3.9]) the following strengthening of Proposition 6.1.11 that takes into account crossings at $\pm\infty$:

Proposition 6.1.13 *Let $(\tilde{\theta}_i)_{i\in I}$ and $(\tilde{\theta}'_i)_{i\in I}$ be two distinct minimizing chains. Assume furthermore that the sequence $|\tilde{\theta}_{i+1} - \tilde{\theta}_i|$ is bounded. Then $(\tilde{\theta}_i)_{i\in I}$ and $(\tilde{\theta}'_i)_{i\in I}$ cross at most once, except possibly at both ends of I.*

In the latter case, both chains are maximal minimizing chains.

Remark 6.1.14 Of course, in the previous Proposition, we insist that possible crossings at $\pm\infty$ do count.

A consequence of our forthcoming analysis will be that the hypothesis concerning the boundedness of $|\tilde{\theta}_{i+1} - \tilde{\theta}_i|$ is in fact automatically verified.

This section ends with a fundamental property of minimizing sequences. It is by no means a direct consequence of the previous stated facts. The proof is quite tricky and we refer the interested reader to [30, Theorem 3.15] for details.

Theorem 6.1.15 (Aubry–Mather) *Let $(\tilde{\theta}_i)_{i\in\mathbb{Z}}$ be a minimizing sequence. Then the following hold:*

1. *For all $(a,b)\in\mathbb{Z}^2$, $(\tilde{\theta}_i)_{i\in\mathbb{Z}}$ and $(\tilde{\theta}_{i-a}+b)_{i\in\mathbb{Z}}$ do not cross.*
2. *There exists a homeomorphism $\tilde{g}:\mathbb{R}\to\mathbb{R}$ verifying $\tilde{g}(x+1)=\tilde{g}(x)+1$ for all $x\in\mathbb{R}$ and such that*[1]
$$\forall i\in\mathbb{Z},\quad \tilde{g}(\tilde{\theta}_i)=\tilde{\theta}_{i+1}.$$
3. *It follows then from Poincaré theory that there exists a real number $\rho\in\mathbb{R}$ called rotation number such that*
$$\forall i\in\mathbb{Z},\quad |\tilde{\theta}_i-\tilde{\theta}_0-i\rho|<1.$$
In particular, $\lim_{|i|\to+\infty}\frac{\tilde{\theta}_i}{i}=\rho.$

[1] g is the lift of an orientation preserving circle homeomorphism.

6.2 Examples and Moser's Theorem

6.2.1 *Notions of Integrability*

Our first family of examples are called *integrable ECTMs*. They are of the following form:

Example 6.2.1 Let $\rho : \mathbb{R} \to \mathbb{R}$ be an increasing diffeomorphism. Then the map $f_\rho : (\theta, r) \mapsto (\theta + \rho(r), r)$ is an ECTM.

Such maps have a particularly simple dynamics. Indeed, for all $r \in \mathbb{R}$ if we denote by $C_r = \{(\theta, r), \quad \theta \in \mathbb{T}^1\}$ the canonical circle of height r, then C_r is invariant by f_ρ and the dynamics of f_ρ restricted to C_r is a rotation of angle $\rho(r)$ (mod 1). One checks that f_ρ is the time 1 map of the Tonelli Hamiltonian flow of $H_\rho : (\theta, r) \mapsto \varrho(r)$ where ϱ is a primitive of ρ.

By extension, let us introduce several classes of ECTM exhibiting similar dynamical features.

Definition 6.2.2 Let $f : \mathbb{A} \to \mathbb{A}$ be an ECTM. We say f is integrable if there exists a C^1 area preserving diffeomorphism $\Phi : \mathbb{A} \to \mathbb{A}$ and $\rho : \mathbb{R} \to \mathbb{R}$ such that $f = \Phi^{-1} \circ \tilde{f}_\rho \circ \Phi$.

Remark 6.2.3 In the case of an integrable ECTM, the function ρ will automatically be an increasing diffeomorphism. However, note that the previous definition can be extended trivially to any area preserving transformation of the annulus.

In the previous definition, the transformation f is the time 1 map of the Hamiltonian flow[2] of $H_\rho \circ \Phi$.

The following notions are obtained by considering twist maps for which the annulus $\mathbb{A}$ is foliated by invariant circles. Bare in mind that by Birkhoff's Theorem 6.1.3, such circles are automatically Lipschitz graphs. The integrable case corresponds to a classical foliation. The next notions are obtained by weakening the regularity of the invariant foliation. We therefore start by defining what are those non–regular foliations.

Definition 6.2.4 A continuous foliation $\mathcal{F} = \{\mathcal{F}_c, \ c \in \mathbb{R}\}$, (otherwise called lamination) of $\mathbb{A}$ by graphs is defined through a continuous function $\eta : (\theta, c) \mapsto \eta_c(\theta)$ from $\mathbb{A}$ to $\mathbb{R}$ such that

- for all $\theta \in \mathbb{R}$, the map $c \mapsto \eta_c(\theta)$ is a (increasing) homeomorphism of $\mathbb{R}$,
- for all $c \in \mathbb{R}$, $\int_{\mathbb{T}^1} \eta_c(\theta) d\theta = c$.

[2] This is a general fact and proves that the group of Hamiltonian diffeomorphisms is a normal subgroup of the group of symplectomorphisms (see [28, Exercise 7 page 471]).

For $c \in \mathbb{R}$ the circle $\mathcal{F}_c = \big\{\big(\theta, \eta_c(\theta)\big), \quad \theta \in \mathbb{T}^1\big\}$ is the leaf of the continuous foliation at cohomology c. It then follows that $\mathbb{A}$ is the disjoint union of the leaves of the foliation.

Reciprocally, any such function $\eta : (\theta, c) \mapsto \eta_c(\theta)$ from $\mathbb{A}$ to $\mathbb{R}$ defines a continuous foliation of $\mathbb{A}$ by graphs.

Remark 6.2.5 The second point in the definition of η is a normalization condition. It could be dropped to give an equivalent notion. However, it is so convenient we prefer to include it directly in the definition.

For an intermediate regularity, arises naturally the notion of Lipschitz foliation:

Definition 6.2.6 A Lipschitz foliation $\mathcal{F} = \{\mathcal{F}_c, \ c \in \mathbb{R}\}$ of $\mathbb{A}$ by graphs is the data of a continuous foliation with associated function $\eta : (\theta, c) \mapsto \eta_c(\theta)$ from $\mathbb{A}$ to $\mathbb{R}$ such that

$$\exists K > 0, \forall (c_1, c_2) \in \mathbb{R}^2, \forall \theta \in \mathbb{T}^1, \quad \frac{1}{K}|c_1 - c_2| \leqslant |\eta_{c_1}(\theta) - \eta_{c_2}(\theta)| \leqslant K|c_1 - c_2|.$$

It is now at grasp to define weaker notions of integrability for twist maps:

Definition 6.2.7

- An ECTM f of the annulus is C^0–integrable if there exists a continuous foliation, $\mathcal{F} = \{\mathcal{F}_c, \ c \in \mathbb{R}\}$, of $\mathbb{A}$ by graphs, such that for all $c \in \mathbb{R}$, $f(\mathcal{F}_c) = \mathcal{F}_c$.
- An ECTM f of the annulus is Lipschitz integrable if there exists a Lipschitz foliation, $\mathcal{F} = \{\mathcal{F}_c, \ c \in \mathbb{R}\}$, of $\mathbb{A}$ by graphs, such that for all $c \in \mathbb{R}$, $f(\mathcal{F}_c) = \mathcal{F}_c$.

Those various notions of integrability each have dynamical consequences on the underlying ECTM. Such results are presented without proof in the last Sect. 6.7. Research on those notions is quite frustrating though, starting from the fact that it is still conjectural whether those three notions of integrability are different. For instance, there are no known examples of ECTM that are C^0–integrable but not integrable (meaning that all the leaves are smooth).

On the bright side, it is proved in [21, 22] that continuous foliations by graphs that are invariant by an ECTM must satisfy some particular properties. This is used to exhibit foliations that cannot be invariant by an ECTM.

Theorem 6.2.8 *Let $\mathcal{F} = \{\mathcal{F}_c, \ c \in \mathbb{R}\}$ be the foliation associated to the function $\eta_c(\theta) = c + \varepsilon(c)\cos(2\pi\theta)$, where $\varepsilon : \mathbb{R} \to \mathbb{R}$ is a non C^1, Lipschitz, function with Lipschitz constant less than $(2\pi)^{-1}$. Then $\mathcal{F}$ is not invariant by any ECTM.*

Though the full proof of this Theorem goes beyond the scope of the present text, some explanations will be provided at the very end, in Sect. 6.7.

6.2.2 *The Standard Family*

According to Remark 6.1.5 it is very easy to construct twist maps with no particular property. However, let us mention a historically very important family of examples.

Example 6.2.9 For all $\varepsilon \in \mathbb{R}$ let us define the map $F_\varepsilon : \mathbb{A} \to \mathbb{A}$ by

$$\forall(\theta, r) \in \mathbb{A}, \quad F_\varepsilon(\theta, r) = \left(\theta + r - \frac{\varepsilon}{2\pi}\sin(2\pi\theta), r - \frac{\varepsilon}{2\pi}\sin(2\pi\theta)\right).$$

This family serves as a test for the state–of–the–art research on twist maps. On certain aspects, the picture is not glorious. For instance, a conjecture of Sinai is that F_ε has positive metric entropy for all $\varepsilon \neq 0$ (with respect to the Lebesgue measure) even though nobody knows how to prove it even for a single parameter. More on such questions is discussed in works of Berger and Yoccoz [40] and Berger and Turaev [39]. A picture of the dynamics for some arbitrary value of ε is presented in the introduction of [126] and clearly shows that such a dynamics is very rich. On the one hand, for $\varepsilon = 0$ the map F_0 is the most basic example among integrable ECTMs. For ε small, KAM theory applies and many invariant circles with diophantine rotation numbers persist (see [129, 130] and references therein). On the other hand, a theorem of Mather [160] states that for $\varepsilon > \frac{4}{3}$, F_ε has no invariant essential circle.

What happens to the circles when they disappear is a challenging question: what is their regularity, the dynamics on them at the last parameter... After they disappear Aubry–Mather theory provides an answer as to what they become.

6.2.3 *General Twist Maps and a Theorem of Moser*

Amongst other examples and open questions, the nature of possible invariant circles of twist maps is still not well understood. In the recent [29], the authors construct a C^1 ECTM having an invariant circle that is not everywhere differentiable with a minimal (irrational) restricted dynamics on it.[3] This answers partially a question of Arnaud [10] asking whether such C^2 maps exist.

Concerning invariant essential circles on which the dynamics is irrational and not minimal (that of a Denjoy counterexample) examples have been constructed in [129] where a C^1 invariant circle is constructed, and in [15] where a non–differentiable

[3] Recall that a homeomorphism of the circle having an irrational rotation number is either minimal (all orbits are dense) or not. In the first case, it is conjugated to the irrational rotation. In the second case, it is only semi–conjugated to the irrational rotation as there are wandering intervals. We then speak of a Denjoy counterexample. A Theorem of Denjoy states that Denjoy counterexamples cannot be of class C^2.

invariant curve is constructed (thus answering a question of Mather). By Denjoy's Theorem, such an invariant circle cannot be C^2.

Last, concerning the inverse problem, in [6] are provided examples of essential circles that are graphs of Lipschitz maps but cannot be invariant by any ECTM.

Before turning back to weak KAM theory, let us mention an important Theorem of Moser [178] stating that any regular ECTM can be represented by a Hamiltonian function. The drawback is that the latter is not autonomous (i.e. it is time–dependent).

Theorem 6.2.10 (Moser) *Let $f : \mathbb{A} \to \mathbb{A}$ be an ECTM. There exists a C^2 time–dependent Hamiltonian $H : \mathbb{R} \times \mathbb{A} \to \mathbb{R}$ such that*

- *for all $t \in \mathbb{R}$, $H(t, \cdot, \cdot) : \mathbb{A} \to \mathbb{R}$ is a Tonelli Hamiltonian;*
- *for all $(t, \theta, r) \in \mathbb{R} \times \mathbb{A}$, then $H(t+1, \theta, r) = H(t, \theta, r)$;*
- *the ECTM f is the time 1 map φ_0^1 of the Hamiltonian flow of H generated by the equations*

$$\begin{cases} \dot\theta(t) = \partial_r H\big(t, \theta(t), r(t)\big), \\ \dot r(t) = -\partial_\theta H\big(t, \theta(t), r(t)\big). \end{cases} \tag{6.9}$$

Note that in Moser's original article, only the case of a C^∞ ECTM is treated. However, it is stated in the paper that the proof adapts to less regular functions. Moreover, it can be checked that Moser's construction allows to interpolate between Identity and f by twist maps using the family $(\varphi_0^t)_{t\in[0,1]}$. The generating functions $(\widetilde{S}_t)_{t\in(0,1]}$ of those twist maps are given by the relations

$$\tilde\varphi_0^t(\tilde\theta, r) = (\widetilde\Theta, R) \iff \widetilde{S}_t(\tilde\theta, \widetilde\Theta) = \int_0^t \widetilde{L}\big(s, \gamma_{(\tilde\theta,r)}(s), \dot\gamma_{(\tilde\theta,r)}(s)\big)\, \mathrm{d}s,$$

where the Lagrangian function $L : \mathbb{R} \times \mathbb{A} \to \mathbb{R}$ is defined as previously by

$$\forall (t, \theta, v) \in \mathbb{R} \times \mathbb{A}, \quad L(t, \theta, v) = \sup_{r\in\mathbb{R}} rv - H(t, \theta, r),$$

and the curve $\gamma_{(\tilde\theta,r)}$ is defined by

$$\forall s \in \mathbb{R}, \quad \gamma_{(\tilde\theta,r)}(s) = \pi_1 \circ \tilde\varphi_0^s(\tilde\theta, r).$$

Here, the generating function is given by the Lagrangian (minimal) action.

6.3 Weak KAM for Twist Maps

We now turn to the study of the Lax–Oleinik semigroup and weak KAM solutions for an ECTM, $f : \mathbb{A} \to \mathbb{R}$. As was already apparent in the previous Chapter dedicated to examples, there is actually a one parameter family of semigroups, indexed by the cohomology. The underlying metric space here is $X = \mathbb{T}^1$ that is obviously compact. We will precisely build a 1–parameter family of cost functions using that the first cohomology group $H^1(\mathbb{T}^1, \mathbb{R})$ is isomorphic to $\mathbb{R}$. This appears in the early works of Mather on twist maps that later led him to define the α function in a higher dimensional Lagrangian setting [165]. The latter idea was also adapted in [201] on more general metric spaces.

As previously, S is a generating function of f, $\widetilde{S} : \mathbb{R}^2 \to \mathbb{R}$ is the lift of S and we choose $\tilde{f} = (\widetilde{\Theta}, R) : \mathbb{R}^2 \to \mathbb{R}^2$ a lift of f. The canonical projection is denoted by $\pi : \mathbb{R} \to \mathbb{T}^1$.

Definition 6.3.1 Let $c \in \mathbb{R}$ that we will refer to as a cohomology class. The cost function $S^c : \mathbb{T}^1 \times \mathbb{T}^1 \to \mathbb{R}$ is defined by

$$\forall (\theta, \Theta) \in \mathbb{T}^1 \times \mathbb{T}^1, \quad S^c(\theta, \Theta) = \inf_{\substack{\pi(\tilde{\theta})=\theta \\ \pi(\widetilde{\Theta})=\Theta}} \widetilde{S}(\tilde{\theta}, \widetilde{\Theta}) + c(\tilde{\theta} - \widetilde{\Theta}).$$

Remark 6.3.2

- In the previous definition, using the translation invariance of $\widetilde{S}$, it is also possible to fix a $\tilde{\theta} \in [-1, 1]$ such that $\pi(\tilde{\theta}) = \theta$ and take the infimum solely on $\widetilde{\Theta}$.
- As $\widetilde{S}$ is superlinear (6.5) it is easily seen that the previous infimum is actually a minimum. Moreover, if as asserted, we restrict to $\tilde{\theta} \in [-1, 1]$ then there exists $K > 0$ (depending on c) such that the minimum on $\widetilde{\Theta}$ can be restricted to $\widetilde{\Theta} \in [-K, K]$.
- Given $\theta \in \mathbb{T}^1$ and $\tilde{\theta} \in \mathbb{R}$ such that $\pi(\tilde{\theta}) = \theta$, the derivative $\partial_2 S^c(\theta, \Theta)$ exists if and only if there exists a unique $\widetilde{\Theta}_0$ realizing the minimum $S^c(\theta, \Theta) = \widetilde{S}(\tilde{\theta}, \widetilde{\Theta}_0) + c(\tilde{\theta} - \widetilde{\Theta}_0)$. In such a case, $\partial_2 S^c(\theta, \Theta) = \partial_2 S(\tilde{\theta}, \widetilde{\Theta}_0) - c$.
- Similarly, given $\Theta \in \mathbb{T}^1$ and $\widetilde{\Theta} \in \mathbb{R}$ such that $\pi(\widetilde{\Theta}) = \Theta$, the derivative $\partial_1 S^c(\theta, \Theta)$ exists if and only if there exists a unique $\tilde{\theta}_0$ realizing the minimum $S^c(\theta, \Theta) = \widetilde{S}(\tilde{\theta}_0, \widetilde{\Theta}) + c(\tilde{\theta}_0 - \widetilde{\Theta})$. In such a case, $\partial_1 S^c(\theta, \Theta) = \partial_1 \widetilde{S}(\tilde{\theta}_0, \widetilde{\Theta}) + c$.
- The previous points and the fact that f is a twist map yields that S^c verifies the left and right twist conditions as defined in Definition 3.5.2 (see also [201, Proposition 6.4]).
- From the previous points, the same strange regularity property observed for costs coming from Tonelli Lagrangians (see discussion following Proposition 3.6.14) is brought to light: given $(\theta, \Theta) \in \mathbb{T}^1 \times \mathbb{T}^1$, the following are equivalent
  - $\partial_1 S^c(\theta, \Theta)$ exists;
  - $\partial_2 S^c(\theta, \Theta)$ exists.

With those facts and bearing in mind that an infimum of a compact family of C^2 functions is semiconcave, it follows that

Proposition 6.3.3 *The function S^c is locally semiconcave. More precisely, any of its lifts to $\mathbb{R}^2$ is semiconcave.*

In particular, if S^c admits any partial derivative at points (θ, Θ) then S^c is differentiable at (θ, Θ).

From the previous points and the semiconcavity of S^c (see the proof of Proposition 3.4.11 and the following Remark 3.4.12), it follows that:

Proposition 6.3.4 *If $(\theta_k)_{k\in I}$ is a minimizing sequence or chain (with at least 3 points) for S^c then all the derivatives $\partial_i S^c(\theta_k, \theta_{k+1})$, $i \in \{1, 2\}$ and $(k, k+1) \in I \times I$, exist. If $k_0 \in I$ and $\tilde{\theta}_{k_0}$ is a lift of θ_{k_0} then there exists a unique (minimizing) chain $(\tilde{\theta}_k)_{k\in I}$ such that*

$$\forall k < k', \quad \sum_{\ell=k}^{k'-1} S^c(\theta_\ell, \theta_{\ell+1}) = \sum_{\ell=k}^{k'-1} \widetilde{S}(\tilde{\theta}_\ell, \tilde{\theta}_{\ell+1}) + c(\tilde{\theta}_k - \tilde{\theta}_{k'}).$$

As is now customary, setting then

$$r_\ell = -\partial_1 \widetilde{S}(\tilde{\theta}_\ell, \tilde{\theta}_{\ell+1}) = \partial_2 \widetilde{S}(\tilde{\theta}_{\ell-1}, \tilde{\theta}_\ell) = c - \partial_1 S^c(\theta_\ell, \theta_{\ell+1}) = c + \partial_2 S^c(\theta_{\ell-1}, \theta_\ell),$$

then $(\theta_k, r_k)_{k\in I}$ is a piece of orbit of f and $(\tilde{\theta}_k, r_k)_{k\in I}$ is a piece of orbit of $\tilde{f}$.

We will now focus our attention on the negative Lax–Oleinik semigroups associated to S^c and gather results previously proven.

Definition 6.3.5 Given $c \in \mathbb{R}$, we define the operator T^c which, to a bounded function $u : \mathbb{T}^1 \to \mathbb{R}$, associates the function

$$T^c u : \Theta \in \mathbb{T}^1 \mapsto \inf_{\theta\in\mathbb{T}^1} u(\theta) + S^c(\theta, \Theta).$$

Equivalently, recalling that two infimums commute, if $\tilde{u} : \mathbb{R} \to \mathbb{R}$ is the lift of u then the lift of $T^c u$ is given by the relation

$$\forall \widetilde{\Theta} \in \mathbb{R}, \quad \widetilde{T^c} \tilde{u}(\widetilde{\Theta}) = \inf_{\tilde{\theta}\in\mathbb{R}} \tilde{u}(\tilde{\theta}) + \widetilde{S}(\tilde{\theta}, \widetilde{\Theta}) + c(\tilde{\theta} - \widetilde{\Theta}).$$

The weak KAM Theorem then states:

Theorem 6.3.6 *For all $c \in \mathbb{R}$ there exists a unique constant $\alpha(c)$ for which the equation*

$$u = T^c u + \alpha(c) \tag{6.10}$$

admits solutions $u^c : \mathbb{T}^1 \to \mathbb{R}$*. Such a solution (or its lift* $\tilde{u}_c : \mathbb{R} \to \mathbb{R}$*) will be called a weak KAM solution at cohomology* c*.*

The function $\alpha : \mathbb{R} \to \mathbb{R}$ is called *Mather's* α *function.*

Proposition 6.3.7 *Mather's* α *function is convex and superlinear.*

Proof One easily reconstructs from the definition of S^c and Proposition 2.2.7 that if $c \in \mathbb{R}$, then $\alpha(c)$ is the least constant $a \in \mathbb{R}$ such that there exists a 1–periodic function $\tilde{v} : \mathbb{R} \to \mathbb{R}$ verifying

$$\forall(\tilde{\theta}, \tilde{\theta}') \in \mathbb{R} \times \mathbb{R}, \quad \tilde{v}(\tilde{\theta}') - \tilde{v}(\tilde{\theta}) \leqslant \widetilde{S}(\tilde{\theta}, \tilde{\theta}') + c(\tilde{\theta} - \tilde{\theta}') + a.$$

If now c_1 and c_2 are real numbers and $\tilde{u}_1$ and $\tilde{u}_2$ are lifts of weak KAM solutions at the corresponding cohomology classes, if $t \in (0, 1)$ one infers that, setting $\tilde{u}_t = t\tilde{u}_1 + (1-t)\tilde{u}_2$,

$$\begin{aligned} \forall(\tilde{\theta}, \tilde{\theta}') \in \mathbb{R} \times \mathbb{R}, \quad & \tilde{u}_t(\tilde{\theta}') - \tilde{u}_t(\tilde{\theta}) \\ & \leqslant \widetilde{S}(\tilde{\theta}, \tilde{\theta}') + (tc_1 + (1-t)c_2)(\tilde{\theta} - \tilde{\theta}') + t\alpha(c_1) + (1-t)\alpha(c_2). \end{aligned}$$

It follows that $\alpha(tc_1 + (1-t)c_2) \leqslant t\alpha(c_1) + (1-t)\alpha(c_2)$ and the convexity is proved.

For superlinearity, let $k > 0$ be an integer. If $c \in \mathbb{R}$, let $u_c : \mathbb{T}^1 \to \mathbb{R}$ be a weak KAM solution at cohomology c. If $\tilde{u}_c$ is its lift, by periodicity, we infer that

$$0 = \tilde{u}_c(k) - \tilde{u}_c(0) \leqslant \widetilde{S}(0, k) - ck + \alpha(c),$$

$$0 = \tilde{u}_c(-k) - \tilde{u}_c(0) \leqslant \widetilde{S}(0, -k) + ck + \alpha(c).$$

From those inequalities, we infer that, setting $C_k = \max\big(\widetilde{S}(0, k), \widetilde{S}(0, -k)\big)$,

$$\forall c \in \mathbb{R}, \quad \alpha(c) \geqslant k|c| - C_k,$$

from which it follows that $\lim\limits_{|c| \to +\infty} \dfrac{\alpha(c)}{|c|} = +\infty.$

□

Informations obtained from Propositions 3.4.11, 6.1.7, and Remark 6.3.2 are gathered in the next Theorem (keeping in mind that a weak KAM solution is locally semiconcave):

Theorem 6.3.8 *Let* $c \in \mathbb{R}$*,* $u_c : \mathbb{T}^1 \to \mathbb{R}$ *be a weak KAM solution at cohomology* c *and* $\tilde{u}_c : \mathbb{R} \to \mathbb{R}$ *its lift.*

1. *For all* $\theta_0 \in \mathbb{T}^1$, *there exists* $(\theta_k^c)_{k\leqslant 0}$ *such that* $\theta_0 = \theta_0^c$ *and*

$$\forall k < 0,\ \ u_c(\theta_0) = u_c(\theta_k^c) + \sum_{i=k}^{-1} S^c(\theta_i^c, \theta_{i+1}^c) + |k|\alpha(c).$$

2. *The previous sequence may not be unique but it is uniquely determined by* $(\theta_{-1}^c, \theta_0)$. *Moreover, setting for* $k \leqslant 0$, $r_k = c + \partial_2 S^c(\theta_{k-1}^c, \theta_k^c)$ *(that exists) then* $(\theta_k^c, r_k)_{k\leqslant 0}$ *is a piece of orbit of* f.
3. *For all* $k < 0$, u_c *is derivable at* θ_k^c *and* $c+u_c'(\theta_k^c) = r_k$. *Moreover,* $r_0 \in \partial^+ u_c(\theta_0)$ *and* u_c *is derivable at* θ_0 *if and only if the sequence* $(\theta_k^c)_{k\leqslant 0}$ *is unique.*
4. *It follows that* $f^{-1}\left(\overline{\mathcal{G}(c+u_c')}\right) \subset \mathcal{G}(c+u_c')$ *where* $\mathcal{G}(c+u_c')$ *is the set of* $\big(\theta, c+u_c'(\theta)\big)$ *for* $\theta \in \mathbb{T}^1$ *such that* $u_c'(\theta)$ *exists.*[4] *Moreover, if* $(\theta_0, r_0) \in \overline{\mathcal{G}(c+u_c')}$ *and for* $k \leqslant 0$, $(\theta_k, r_k) = f^k(\theta_0, r_0)$, *then*

$$\forall k < 0,\ \ u_c(\theta_0) = u_c(\theta_k) + \sum_{i=k}^{-1} S^c(\theta_i, \theta_{i+1}) + |k|\alpha(c).$$

5. *Given a sequence* $(\theta_k^c)_{k\leqslant 0}$ *as above and* $\tilde\theta_0 \in \mathbb{R}$ *a lift of* θ_0, *there exists a unique* $(\tilde\theta_k^c)_{k\leqslant 0}$ *that projects on* $(\theta_k^c)_{k\leqslant 0}$ *and such that*

$$\forall k < 0,\ \ \tilde u_c(\tilde\theta_0) = \tilde u_c(\tilde\theta_k^c) + \sum_{i=k}^{-1} \widetilde{S}(\tilde\theta_i^c, \tilde\theta_{i+1}^c) + c(\tilde\theta_k^c - \tilde\theta_0^c) + |k|\alpha(c).$$

6. *With the previous notations for* $k \leqslant 0$, $r_k = \partial_2 \widetilde{S}(\tilde\theta_{k-1}^c, \tilde\theta_k^c)$ *and* $(\tilde\theta_k^c, r_k)_{k\leqslant 0}$ *is a piece of orbit of* $\tilde f$.
7. *For all* $k < 0$, $\tilde u_c$ *is derivable at* $\tilde\theta_k^c$ *and* $c+\tilde u_c'(\tilde\theta_k^c) = r_k$. *Moreover,* $r_0 \in \partial^+ \tilde u_c(\tilde\theta_0)$ *and* $\tilde u_c$ *is derivable at* $\tilde\theta_0$ *if and only if the sequence* $(\tilde\theta_k^c)_{k\leqslant 0}$ *is unique.*
8. *It follows that* $\tilde f^{-1}\left(\overline{\mathcal{G}(c+\tilde u_c')}\right) \subset \mathcal{G}(c+\tilde u_c')$ *where* $\mathcal{G}(c+\tilde u_c')$ *is the set of* $\big(\theta, c+\tilde u_c'(\theta)\big)$ *for* $\tilde\theta \in \mathbb{R}$ *such that* $\tilde u_c'(\tilde\theta)$ *exists. Finally, if* $(\tilde\theta_0, r_0) \in \overline{\mathcal{G}(c+\tilde u_c')}$ *and for* $k \leqslant 0$, $(\tilde\theta_k, r_k) = \tilde f^k(\theta_0, r_0)$, *then*

$$\forall k < 0,\ \ \tilde u_c(\tilde\theta_0) = \tilde u_c(\tilde\theta_k) + \sum_{i=k}^{-1} \widetilde{S}(\tilde\theta_i, \tilde\theta_{i+1}) + c(\tilde\theta_k - \tilde\theta_0) + |k|\alpha(c).$$

---

[4] In the previous inclusion, the fact that we can take a closure on the left hand side is because a limit of calibrating sequences for u_c is still calibrating.

Remark 6.3.9 Being in dimension 1, concave functions have very strong derivability properties that semiconcave functions inherit from. It follows that if $v : \mathbb{T}^1 \to \mathbb{R}$ is a semiconcave function, then it admits at all $\theta \in \mathbb{T}^1$ a left derivative $v'_-(\theta)$ and a right derivative $v'_+(\theta)$ that verify $v'_-(\theta) \geqslant v'_+(\theta)$. Its superdifferential at θ is then the segment $\partial^+ v(\theta) = [v'_+(\theta), v'_-(\theta)]$. Finally, using the previous notation, if $c \in \mathbb{R}$ then

$$\overline{\mathcal{G}(c + v')} = \bigcup_{\theta \in \mathbb{R}} \{\theta\} \times \{c + v'_+(\theta), c + v'_-(\theta)\}.$$

We introduce the notion of full pseudograph that is the graph of the superdifferential of a semiconcave function and appears in various weak KAM related works [12, 43]:

Definition 6.3.10 Let $v : \mathbb{T}^1 \to \mathbb{R}$ be a semiconcave function. Then its full pseudograph is

$$\mathcal{PG}(v') = \bigcup_{\theta \in \mathbb{T}^1} \{\theta\} \times \partial^+ v(\theta) = \bigcup_{\theta \in \mathbb{T}^1} \{\theta\} \times [v'_+(\theta), v'_-(\theta)].$$

A related notation: for $c \in \mathbb{R}$,

$$\mathcal{PG}(c + v') = \bigcup_{\theta \in \mathbb{T}^1} \{\theta\} \times \big(c + \partial^+ v(\theta)\big) = \bigcup_{\theta \in \mathbb{T}^1} \{\theta\} \times [c + v'_+(\theta), c + v'_-(\theta)].$$

We will use similar notations for semiconcave functions on $\mathbb{R}$.

A beautiful theorem due to Marie-Claude Arnaud (using weak KAM methods) yields:

Proposition 6.3.11 *Let $v : \mathbb{T}^1 \to \mathbb{R}$ be a semiconcave function. Then its full pseudograph $\mathcal{PG}(v')$ is a Lipschitz manifold. In the present case it is a Lipschitz essential circle (meaning it separates the annulus in two unbounded connected components).*

We are now ready to state a first result on the interplay between Lax–Oleinik and the non–crossing lemma:

Lemma 6.3.12 *Let $v : \mathbb{T}^1 \to \mathbb{R}$ be a continuous function, $c \in \mathbb{R}$ and $\tilde{\theta}_1 < \tilde{\theta}_2$ two real numbers.*

1. *Assume that $\tilde{\theta}'_1$ and $\tilde{\theta}'_2$ verify for $i \in \{1, 2\}$,*

$$\widetilde{T}^c \tilde{v}(\tilde{\theta}_i) = \min_{\tilde{\theta}' \in \mathbb{R}} \big(\tilde{v}(\tilde{\theta}') + \widetilde{S}(\tilde{\theta}', \tilde{\theta}_i) + c(\tilde{\theta}' - \tilde{\theta}_i)\big) = \tilde{v}(\tilde{\theta}'_i) + \widetilde{S}(\tilde{\theta}'_i, \tilde{\theta}_i) + c(\tilde{\theta}'_i - \tilde{\theta}_i).$$

   Then $\tilde{\theta}'_1 \leqslant \tilde{\theta}'_2$.
2. *If moreover v is semiconcave, then $\tilde{\theta}'_1 < \tilde{\theta}'_2$.*

Proof Let us argue by contradiction. Then by Proposition 6.1.10 the following holds:

$$\tilde{v}(\tilde{\theta}_1') + \widetilde{S}(\tilde{\theta}_1', \tilde{\theta}_1) + c(\tilde{\theta}_1' - \tilde{\theta}_1) + \tilde{v}(\tilde{\theta}_2') + \widetilde{S}(\tilde{\theta}_2', \tilde{\theta}_2) + c(\tilde{\theta}_2' - \tilde{\theta}_2) >$$
$$> \tilde{v}(\tilde{\theta}_2') + \widetilde{S}(\tilde{\theta}_2', \tilde{\theta}_1) + c(\tilde{\theta}_2' - \tilde{\theta}_1) + \tilde{v}(\tilde{\theta}_1') + \widetilde{S}(\tilde{\theta}_1', \tilde{\theta}_2) + c(\tilde{\theta}_1' - \tilde{\theta}_2).$$

We infer that at least one of the two inequalities

$$\tilde{v}(\tilde{\theta}_1') + \widetilde{S}(\tilde{\theta}_1', \tilde{\theta}_1) + c(\tilde{\theta}_1' - \tilde{\theta}_1) > \tilde{v}(\tilde{\theta}_2') + \widetilde{S}(\tilde{\theta}_2', \tilde{\theta}_1) + c(\tilde{\theta}_2' - \tilde{\theta}_1),$$

$$\tilde{v}(\tilde{\theta}_2') + \widetilde{S}(\tilde{\theta}_2', \tilde{\theta}_2) + c(\tilde{\theta}_2' - \tilde{\theta}_2) > \tilde{v}(\tilde{\theta}_1') + \widetilde{S}(\tilde{\theta}_1', \tilde{\theta}_2) + c(\tilde{\theta}_1' - \tilde{\theta}_2),$$

is valid that is a contradiction.

To prove the second item, we recall that thanks to Proposition 3.4.11, if $\tilde{v}$ is semiconcave, then it is derivable both at $\tilde{\theta}_1'$ and $\tilde{\theta}_2'$ and $\pi_1 \circ \tilde{f}\big(\tilde{\theta}_i', c + \tilde{v}'(\tilde{\theta}_i')\big) = \tilde{\theta}_i$. As $\tilde{\theta}_1 \neq \tilde{\theta}_2$, necessarily $\tilde{\theta}_1' \neq \tilde{\theta}_2'$.

□

An interesting Corollary, reminiscent of Bernard's Theorem 3.6.11 and that will be needed later is:

Corollary 6.3.13 *Let $c \in \mathbb{R}$ and $u_c : \mathbb{T}^1 \to \mathbb{R}$ be a weak KAM solution at cohomology c. Then $T^{c+}u_c$ is a C^1 function (where T^{c+} is the positive Lax–Oleinik semigroup given by Definition 2.3.1, associated to S^c) and $\mathcal{G}\big(c + (T^{c+}u_c)'\big) = f^{-1}\big(\mathcal{PG}(c + u_c')\big)$.*

Proof We start by proving that $f^{-1}\big(\mathcal{PG}(c + u_c')\big)$ is the graph of a continuous function. By Proposition 6.3.11, there exists a Lipschitz embedding $\gamma : \mathbb{T}^1 \to \mathbb{A}$ such that $\gamma(\mathbb{T}^1) = \mathcal{PG}(c + u_c')$. Denote by $\tilde{\gamma} : \mathbb{R} \to \mathbb{R} \times \mathbb{R}$ a lift of γ and set $\tilde{\gamma} = (\tilde{\gamma}_1, \tilde{\gamma}_2)$ the coordinates of $\tilde{\gamma}$. As u_c' is semiconcave, up to reversing the time parametrization of γ we may assume that $\tilde{\gamma}_1$ is non–decreasing and it follows that $\tilde{\gamma}_2$ is decreasing on intervals where $\tilde{\gamma}_1$ is constant. We now define for all $t \in \mathbb{R}$, $\big(\Gamma_1(t), \Gamma_2(t)\big) = f^{-1}\big(\tilde{\gamma}_1(t), \tilde{\gamma}_2(t)\big)$. Let us establish that Γ_1 is increasing that will imply our point.

Let $t_1 < t_2$.

- Assume for a start that $\tilde{\gamma}_1(t_1) = \tilde{\gamma}_1(t_2)$. It follows that $\tilde{\gamma}_2(t_1) > \tilde{\gamma}_2(t_2)$ and by the twist condition, we deduce that $\Gamma_1(t_1) < \Gamma_1(t_2)$.
- For the remaining case, $\tilde{\gamma}_1(t_1) < \tilde{\gamma}_1(t_2)$, define now $T_1 = \max\big(t \geqslant t_1,\ \tilde{\gamma}_1(t_1) = \tilde{\gamma}_1(t)\big)$ and $T_2 = \min\big(t \leqslant t_2,\ \tilde{\gamma}_1(t_2) = \tilde{\gamma}_1(t)\big)$. It follows that $t_1 \leqslant T_1 < T_2 \leqslant t_2$ and that

$$\tilde{\gamma}_2(T_1) = c + u_{c+}'\big(\tilde{\gamma}_1(t_1)\big), \quad \tilde{\gamma}_2(T_2) = c + u_{c-}'\big(\tilde{\gamma}_1(t_2)\big). \tag{6.11}$$

As $\tilde{\gamma}_1(t_1) = \tilde{\gamma}_1(T_1)$ and $\tilde{\gamma}_1(t_2) = \tilde{\gamma}_1(T_2)$ we obtain from the first case that $\Gamma_1(t_1) \leqslant \Gamma_1(T_1)$ and $\Gamma_1(T_2) \leqslant \Gamma_1(t_2)$.

Then by (6.11) and Theorem 6.3.8, we deduce that for $i \in \{1, 2\}$,

$$\widetilde{T}^c \tilde{u}_c\big(\tilde{\gamma}_1(t_i)\big) = \tilde{u}_c\big(\Gamma_1(T_i)\big) + \widetilde{S}\big(\Gamma_1(T_i), \tilde{\gamma}_1(t_i)\big) + c\big(\Gamma_1(T_i) - \tilde{\gamma}_1(t_i)\big).$$

By Lemma 6.3.12 we obtain that $\Gamma_1(T_1) < \Gamma_1(T_2)$ and finally

$$\Gamma_1(t_1) \leqslant \Gamma_1(T_1) < \Gamma_1(T_2) \leqslant \Gamma_1(t_2).$$

We now turn to the interpretation in terms of positive Lax–Oleinik semigroup. To this end, we use the analogues for T^{c+} of the results established for T^c, without proofs. Note that $T^{c+}u_c$ is a semiconvex function. Let $\theta \in \mathbb{T}^1$ and $\Theta \in \mathbb{T}^1$ such that $T^{c+}u_c(\theta) = u_c(\Theta) - S^c(\theta, \Theta)$. Then S^c is differentiable at (θ, Θ). By setting $R = c + \partial_2 S^c(\theta, \Theta)$ and $r = c - \partial_1 S^c(\theta, \Theta)$,

- $f(\theta, r) = (\Theta, R)$,
- $(\Theta, R) \in \mathcal{PG}(c + u_c')$,
- $r - c \in \partial^- T^{c+}u_c(\theta)$.

As we have established that $f^{-1}\big(\mathcal{PG}(c+u_c')\big)$ is the graph of a continuous function, there is a unique $(\Theta, R) \in \mathcal{PG}(c + u_c')$ such that $\pi_1 \circ f^{-1}(\Theta, R) = \theta$. It follows that Θ $\big($realizing equality in the definition of $T^{c+}u_c(\theta)\big)$ is unique and that $T^{c+}u_c$ is derivable at θ. As this holds for all θ and by semiconvexity, $T^{c+}u_c$ is indeed C^1. Finally, as $\mathcal{G}\big(c + (T^{c+}u_c)'\big) \subset f^{-1}\big(\mathcal{PG}(c + u_c')\big)$ and since both sets are graphs, they are equal.

□

Remark 6.3.14 The previous result was new when this book was first written and it has since been improved in [98] where it is proved that $T^{c+}u_c$ is a $C^{1,1}$.

Definition 6.3.15 Given $c \in \mathbb{R}$ we will denote by

- $\mathcal{A}_c \subset \mathbb{T}^1$ the projected Aubry set,
- $\widehat{\mathcal{A}}_c \subset \mathbb{T}^1 \times \mathbb{T}^1$ the 2-Aubry set,
- $\widetilde{\mathcal{A}}_c \subset (\mathbb{T}^1)^{\mathbb{Z}}$ the Aubry set,

all three associated to the cost $\widetilde{S}_c$.

We similarly denote by $\mathcal{A}_c^* \subset \mathbb{A} = \mathbb{T}^1 \times \mathbb{R}$ the set given by Proposition 3.5.1 associated to the cost $\widetilde{S}_c$ that is also referred to as Aubry set.

We will denote by $\mathfrak{A}_c$ and $\mathfrak{A}_c^*$ the lifts of $\mathcal{A}_c$ and $\mathcal{A}_c^*$ to respectively $\mathbb{R}$ and $\mathbb{R} \times \mathbb{R}$ that we will also refer to as projected Aubry set and Aubry set.

As sequences in $\widetilde{\mathcal{A}}_c$ are minimizing for S^c we may apply Proposition 6.3.4 to obtain

Proposition 6.3.16 *Let $(\theta_i)_{i\in\mathbb{Z}} \in \widetilde{\mathcal{A}}_c$ and $\tilde{\theta}_0 \in \mathbb{R}$ a lift of θ_0. Then there exists a unique $(\tilde{\theta}_i)_{i\in\mathbb{Z}} \in \mathbb{R}^{\mathbb{Z}}$ such that*

$$\forall k < k', \quad \sum_{\ell=k}^{k'-1} S^c(\theta_\ell, \theta_{\ell+1}) = \sum_{\ell=k}^{k'-1} \widetilde{S}(\tilde{\theta}_\ell, \tilde{\theta}_{\ell+1}) + c(\tilde{\theta}_k - \tilde{\theta}_{k'}).$$

More precisely, if $k \in \mathbb{Z}$, $\tilde{\theta}_k = \pi_1 \circ \tilde{f}^k(\tilde{\theta}_0, r_0)$ where

$$r_0 = -\partial_1 \widetilde{S}(\tilde{\theta}_0, \tilde{\theta}_1) = \partial_2 \widetilde{S}(\tilde{\theta}_{-1}, \tilde{\theta}_0) = c - \partial_1 S^c(\theta_0, \theta_1) = c + \partial_2 S^c(\theta_{-1}, \theta_0).$$

Definition 6.3.17 We denote by $\widetilde{\mathfrak{A}}_c \subset \mathbb{R}^{\mathbb{Z}}$ the set of sequences $(\tilde{\theta}_i)_{i\in\mathbb{Z}} \in \mathbb{R}^{\mathbb{Z}}$ given by the previous proposition.

We denote by $\widehat{\mathfrak{A}}_c \subset \mathbb{R}^2$ the set of pairs $(\tilde{\theta}_0, \tilde{\theta}_1)$ for $(\tilde{\theta}_i)_{i\in\mathbb{Z}} \in \widetilde{\mathfrak{A}}_c$.

Remark 6.3.18

- All canonical projections from respectively $\widehat{\mathcal{A}}_c$, $\widetilde{\mathcal{A}}_c$ and $\mathcal{A}_c^*$ to $\mathcal{A}_c$ are bi–Lipschitz homeomorphisms.
- All canonical projections from respectively $\widehat{\mathfrak{A}}_c$, $\widetilde{\mathfrak{A}}_c$ and $\mathfrak{A}_c^*$ to $\mathfrak{A}_c$ are bi–Lipschitz homeomorphisms.
- The sets $\mathfrak{A}_c$ and $\mathfrak{A}_c^*$ are respectively invariant by horizontal translations $\tilde{\theta} \mapsto \tilde{\theta}+1$ and $(\tilde{\theta}, r) \mapsto (\tilde{\theta}+1, r)$.
- The set $(0, c) + \mathcal{A}_c^* = \{(\theta, c + r),\ \ (\theta, r) \in \mathcal{A}_c^*\}$ is invariant by f and the set $(0, c) + \mathfrak{A}_c^* = \{(\tilde{\theta}, c + r),\ \ (\tilde{\theta}, r) \in \mathfrak{A}_c^*\}$ by $\tilde{f}$.

This last point is proved using that elements in the projected Aubry sets come in minimizing sequences that calibrate weak KAM solutions. Hence it is possible to apply Remark 6.3.2 and Theorem 6.3.8.

We derive the following consequence (that will be improved later in Corollary 6.6.4):

Corollary 6.3.19 *Let $c \in \mathbb{R}$. There exists $\rho(c) \in \mathbb{R}$ such that for all $u_c : \mathbb{T}^1 \to \mathbb{R}$ weak KAM solution at cohomology c, if $(\tilde{\theta}_k)_{k\leqslant 0} \in \mathbb{R}^{\mathbb{Z}_-}$ calibrates $\tilde{u}_c$ then*

$$\forall k \leqslant 0, \quad |\tilde{\theta}_k - \tilde{\theta}_0 - k\rho(c)| < 2.$$

Proof Let $(x_k)_{k\in\mathbb{Z}} \in \widetilde{\mathcal{A}}_c$ that hence calibrates u_c. Let $\tilde{x}_0 \in \mathbb{R}$ such that $\tilde{x}_0 \leqslant \tilde{\theta}_0 < \tilde{x}_0 + 1$. Finally let $(\tilde{x}_k)_{k\in\mathbb{Z}}$ be the only sequence that projects on $(x_k)_{k\in\mathbb{Z}}$ and calibrates $\tilde{u}_c$. Thanks to Theorem 6.1.15, there exists a real number ρ that is independent on u_c such that

$$\forall i \in \mathbb{Z}, \quad |\tilde{x}_i - \tilde{x}_0 - i\rho| < 1.$$

Moreover, by periodicity, the sequences $(\tilde{x}_k + 1)_{k\in\mathbb{Z}}$ also calibrates $\tilde{u}_c$.

If $\tilde{\theta}_0 = \tilde{x}_0$, then $\tilde{\theta}_k = \tilde{x}_k$ for all $k \leqslant 0$. Indeed recall that $\tilde{u}_c'(\tilde{x}_0)$ exists and then a calibrating sequence starting at $\tilde{x}_0$ is unique (see Theorem 6.3.8).

In the remaining case, by applying Lemma 6.3.12 and a straightforward induction, one finds that

$$\forall k \leqslant 0, \quad \tilde{x}_k < \tilde{\theta}_k < \tilde{x}_k + 1,$$

and the result follows.

As all sequences $(x_k')_{k\in\mathbb{Z}} \in \widetilde{\mathcal{A}}_c$ calibrate u_c, it follows that the initial ρ does not depend on the initial choice of $(x_k)_{k\in\mathbb{Z}} \in \widetilde{\mathcal{A}}_c$ (by the previous argument). Finally, as $(x_k)_{k\in\mathbb{Z}} \in \widetilde{\mathcal{A}}_c$ calibrates any other weak KAM solution at cohomology c, the real number ρ does only depend on c, independently of the initially chosen weak KAM solution. □

6.4 Mather Measures

Recall that $\widehat{\mathcal{P}}$ is the set of closed measures on $\mathbb{T}^1 \times \mathbb{T}^1$ (Definition 4.1.1). Then if $c \in \mathbb{R}$, Theorem 4.1.3 stipulates that

$$-\alpha(c) = \min_{\mu\in\widehat{\mathcal{P}}} \int_{\mathbb{T}^1\times\mathbb{T}^1} S_c(\theta, \theta')\, \mathrm{d}\mu(\theta, \theta').$$

Moreover, minimizing Mather measures are those $\mu \in \widehat{\mathcal{P}}$ whose support is included in $\widehat{\mathcal{A}}_c$. We will denote by $\widehat{\mathcal{P}}_c$ the set of such Mather measures at cohomology c. We aim at obtaining analogous notions involving a cost that does not depend on c.

If $\theta \in \mathbb{T}^1$ and $\tilde{\theta}' \in \mathbb{R}$ denote by $\theta + \tilde{\theta}' = \pi(\tilde{\theta} + \tilde{\theta}') \in \mathbb{T}^1$ where $\tilde{\theta} \in \mathbb{R}$ is any lift of θ. Of course, $\theta + \tilde{\theta}'$ does not depend on the choice of $\tilde{\theta}$.

Definition 6.4.1

- Let $\tau : \mathbb{A} \to \mathbb{T}^1$ be defined by $(\theta, r) \mapsto \theta + r$.
- We say a Borel probability measure μ on $\mathbb{A} = \mathbb{T}^1 \times \mathbb{R}$ is closed if it has finite first moment, $\int_{\mathbb{A}} |r|\, \mathrm{d}\mu(\theta, r) < +\infty$ and if $\tau_*\mu = \pi_{1*}\mu$ meaning that for any continuous function $g : \mathbb{T}^1 \to \mathbb{R}$,

$$\int_{\mathbb{A}} g(\theta + r)\, \mathrm{d}\mu(\theta, r) = \int_{\mathbb{A}} g(\theta)\, \mathrm{d}\mu(\theta, r).$$

  The set of closed probability measures on $\mathbb{A}$ is denoted by $\mathcal{P}^*$.
- Given a closed probability measure $\mu \in \mathcal{P}^*$ we define its rotation number, $\rho(\mu) = \int_{\mathbb{A}} r\, \mathrm{d}\mu(\theta, r)$.
- We define $S^* : \mathbb{A} \to \mathbb{R}$ by $S^*(\theta, r) = \widetilde{S}(\tilde{\theta}, \tilde{\theta} + r)$ where $\tilde{\theta} \in \mathbb{R}$ is any lift of θ. Of course, the result does not depend on the choice of $\tilde{\theta}$ by Proposition 6.1.4.

Proposition 6.4.2 *The following holds*

$$-\alpha(c) = \min_{\mu^* \in \mathcal{P}^*} \int_{\mathbb{A}} \left[S^*(\theta, r) - cr\right] \mathrm{d}\mu^*(\theta, r). \tag{6.12}$$

Moreover, a closed measure is minimizing if and only if it is supported on the set of pairs $(\theta, \delta) \in \mathbb{A}$ *such that* $\theta \in \mathcal{A}_c$ *and*

$$\delta = \pi_1 \circ \tilde{f}\left((\pi_{1|\mathfrak{A}_c^*})^{-1}(\tilde{\theta}) + (0, c)\right) - \tilde{\theta} = \pi_1 \circ \tilde{f}(\tilde{\theta}, c + r_\theta) - \tilde{\theta},$$

where $\tilde{\theta} \in \mathbb{R}$ *is any lift of* θ *and* $r_\theta \in \mathbb{R}$ *is the unique real number such that* $(\tilde{\theta}, r_\theta) \in \mathfrak{A}_c^*$.

Proof Let $u : \mathbb{T}^1 \to \mathbb{R}$ be a continuous subsolution for S^c that is strict outside of $\widehat{\mathcal{A}}_c$ (Theorem 2.4.1), meaning that

$$\forall(\theta, \theta') \in \mathbb{T}^1 \times \mathbb{T}^1, \quad u(\theta') - u(\theta) \leqslant S^c(\theta, \theta') + \alpha(c),$$

with strict inequality as soon as $(\theta, \theta') \notin \widehat{\mathcal{A}}_c$. By definition of S^c it follows that if $\tilde{u}$ is a lift of u,

$$\forall(\tilde{\theta}, \tilde{\theta}') \in \mathbb{R} \times \mathbb{R}, \quad \tilde{u}(\tilde{\theta}') - \tilde{u}(\tilde{\theta}) \leqslant \widetilde{S}(\tilde{\theta}, \tilde{\theta}') + c(\tilde{\theta} - \tilde{\theta}') + \alpha(c).$$

This can in turn be written as follows:

$$\forall(\theta, \delta) \in \mathbb{T}^1 \times \mathbb{R}, \quad u(\theta + \delta) - u(\theta) \leqslant S^*(\theta, \delta) - c\delta + \alpha(c).$$

Integrating the previous inequalities against a closed measure $\mu^* \in \mathcal{P}^*$ yields

$$0 = \int_{\mathbb{A}} u(\theta+\delta)\, \mathrm{d}\mu^*(\theta, \delta) - \int_{\mathbb{A}} u(\theta)\, \mathrm{d}\mu^*(\theta, \delta) \leqslant \int_{\mathbb{A}} \left[S^*(\theta, \delta) - c\delta + \alpha(c)\right] \mathrm{d}\mu^*(\theta, \delta)$$

and $-\alpha(c) \leqslant \int_{\mathbb{A}} \left[S^*(\theta, \delta) - c\delta\right] \mathrm{d}\mu^*(\theta, \delta)$.

Moreover, equality holds if and only if μ^* is supported on pairs (θ, δ) such that $u(\theta + \delta) - u(\theta) = S^*(\theta, \delta) - c\delta + \alpha(c)$. As

$$u(\theta + \delta) - u(\theta) \leqslant S^c(\theta, \theta + \delta) + \alpha(c) \leqslant S^*(\theta, \delta) - c\delta + \alpha(c),$$

we deduce that for such $(\theta, \delta) \in \mathrm{supp}(\mu^*)$,

$$u(\theta + \delta) - u(\theta) = S^c(\theta, \theta + \delta) + \alpha(c).$$

In turn, we deduce that $(\theta, \theta+\delta) \in \widehat{\mathcal{A}}_c$, in particular S^c is differentiable at $(\theta, \theta+\delta)$. Then

$$S^c(\theta, \theta+\delta) + \alpha(c) = S^*(\theta, \delta) - c\delta + \alpha(c) = \widetilde{S}(\tilde{\theta}, \tilde{\theta}+\delta) - c\delta + \alpha(c),$$

where $\tilde{\theta}$ is a lift of θ. As weak KAM solutions are calibrated by points of the 2-Aubry set and derivable on the Aubry set, Theorem 6.3.8 (see also Proposition 6.3.16) gives that

$$\tilde{\theta} + r = \pi_1 \circ \tilde{f}(\tilde{\theta}, -\partial_1 \widetilde{S}(\tilde{\theta}, \tilde{\theta}+r))$$

and

$$\big(\tilde{\theta}, -c - \partial_1 \widetilde{S}(\tilde{\theta}, \tilde{\theta}+r)\big) = \big(\tilde{\theta}, -\partial_1 S^c(\theta, \theta+r)\big) \in \mathfrak{A}_c^*.$$

With the notation of the current Proposition, this is rewritten $r_\theta = -\partial_1 S^c(\theta, \theta+r)$.

It remains to prove that such a closed measure realizing equality exists. To that aim let us start from a minimizing Mather measure $\hat{\mu}_c$ such that $-\alpha(c) = \int_{\mathbb{T}^1\times\mathbb{T}^1} S^c(\theta, \theta')\,\mathrm{d}\hat{\mu}_c(\theta, \theta')$, that is henceforth supported on $\widehat{\mathcal{A}}_c$ by Theorem 4.1.3. If $\tilde{\theta} \in \mathfrak{A}_c$ and $\widetilde{\Theta} \in \mathfrak{A}_c$ is the unique element such that $(\tilde{\theta}, \widetilde{\Theta}) \in \widehat{\mathfrak{A}}_c$, we denote $R(\tilde{\theta}) = \widetilde{\Theta} - \tilde{\theta}$. By periodicity, it is immediate that $R(\tilde{\theta})$ only depends on $\theta = \pi(\tilde{\theta}) \in \mathcal{A}_c$ hence we will also refer to it as $R(\theta)$.

Recall that $\tilde{\theta} \mapsto \widetilde{\Theta}$ is the biLipschitz homeomorphism $\pi_2 \circ (\pi_{1|\widehat{\mathfrak{A}}_c})^{-1} : \mathfrak{A}_c \to \mathfrak{A}_c$, as stated in Remark 6.3.18, hence $\theta \mapsto R(\theta)$ is Lipschitz.

We now define a Borel probability measure μ_c^* on $\mathbb{T}^1 \times \mathbb{R}$ by setting, for any continuous function $G : \mathbb{T}^1 \times \mathbb{R} \to \mathbb{R}$,

$$\int_{\mathbb{T}^1\times\mathbb{R}} G(\theta, r)\,\mathrm{d}\mu_c^*(\theta, r) = \int_{\widehat{\mathcal{A}}_c} G\big(\theta, R(\theta)\big)\,\mathrm{d}\hat{\mu}_c(\theta, \theta'). \tag{6.13}$$

As the $R(\theta)$ are uniformly bounded, the measure μ_c^* is compactly supported hence $\int_{\mathbb{A}} |r|\,\mathrm{d}\mu_c^*(\theta, r) < +\infty$. Let us verify it is closed: let $g : \mathbb{T}^1 \to \mathbb{R}$ be a continuous function, then

$$\begin{aligned}
&\int_{\mathbb{T}^1\times\mathbb{R}} \big(g(\theta+r) - g(\theta)\big)\,\mathrm{d}\mu_c^*(\theta, r) \\
&= \int_{\widehat{\mathcal{A}}_c} \big(g(\theta + R(\theta)) - g(\theta)\big)\,\mathrm{d}\hat{\mu}_c(\theta, \theta') = \int_{\widehat{\mathcal{A}}_c} \big(g(\theta') - g(\theta)\big)\,\mathrm{d}\hat{\mu}_c(\theta, \theta') \\
&= \int_{\mathbb{T}^1\times\mathbb{T}^1} \big(g(\theta') - g(\theta)\big)\,\mathrm{d}\hat{\mu}_c(\theta, \theta') = 0,
\end{aligned}$$

where was used that if $(\theta, \theta') \in \widehat{\mathcal{A}}_c$ then $\theta' = \theta + R(\theta)$ and that $\hat{\mu}_c \in \widehat{\mathcal{P}}$ is closed. Hence $\mu_c^* \in \mathcal{P}^*$ is a closed Borel probability measure on $\mathbb{A}$.

To conclude, by definition, μ^* is defined on pairs $\big(\theta, R(\theta)\big)$ where $\theta \in \mathcal{A}_c$ and

$$R(\theta) = \pi_1 \circ \tilde{f}\left((\pi_{1|\mathfrak{A}_c^*})^{-1}(\tilde{\theta}) + (0, c)\right)$$

hence it verifies $-\alpha(c) = \int_{\mathbb{A}} \big[S^*(\theta, r) - cr\big]\, \mathrm{d}\mu_c^*(\theta, r)$ by the first part of the proof. □

Remark 6.4.3

- We will also call minimizing Mather measures (at cohomology c) on $\mathbb{A}$ closed measures μ_c^* verifying $-\alpha(c) = \int_{\mathbb{A}} \big[S^*(\theta, r) - cr\big]\, \mathrm{d}\mu_c^*(\theta, r)$.
- As seen in the previous proof, if $\hat{\mu}_c$ is a minimizing Mather measure on $\mathbb{T}^1 \times \mathbb{T}^1$ (hence supported on $\widehat{\mathcal{A}}_c$) then there is a closed Mather measure μ_c^* on $\mathbb{A}$ that is naturally associated (see (6.13)).
- Reciprocally, if μ_c^* is a closed Mather measure on $\mathbb{A}$, then it is verified the same way that $\hat{\mu}_c = \big((\pi_{|\widehat{\mathcal{A}}_c})^{-1} \circ \pi_1\big)^* \mu_c^*$ is a closed Mather measure on $\mathbb{T}^1 \times \mathbb{T}^1$.
- The two mappings $\hat{\mu}_c \mapsto \mu_c^*$ and $\mu_c^* \mapsto \hat{\mu}_c$ are inverses of one another.

We now relate two notions of rotation number, hence clarifying the terminology.

Proposition 6.4.4 *Let $c \in \mathbb{R}$ be a cohomology class and μ_c^* be a minimizing Mather measure on $\mathbb{A}$. Then $\rho(\mu_c^*) = \rho(c)$ where the first rotation number is provided by Definition 6.4.1 and the second by Corollary 6.3.19.*

Proof By definition, $\rho(\mu_c^*) = \int_{\mathbb{A}} r\, \mathrm{d}\mu_c^*(\theta, r)$. By Proposition 6.4.2 and using notations therein, $\rho(\mu_c^*) = \int_{\mathcal{A}_c \times \mathbb{R}} R(\theta)\, \mathrm{d}\mu_c^*(\theta, r)$. As μ_c^* is closed and supported on pairs of the form $\big(\theta, R(\theta)\big)$, we derive that

$$\rho(\mu_c^*) = \int_{\mathcal{A}_c \times \mathbb{R}} R\big(\theta + R(\theta)\big)\, \mathrm{d}\mu_c^*(\theta, r) = \int_{\mathcal{A}_c \times \mathbb{R}} R_2(\theta)\, \mathrm{d}\mu_c^*(\theta, r).$$

And by induction it follows that for all positive integer $n > 0$,

$$\rho(\mu_c^*) = \int_{\mathcal{A}_c \times \mathbb{R}} R_n(\theta)\, \mathrm{d}\mu_c^*(\theta, r) = \int_{\mathcal{A}_c \times \mathbb{R}} \frac{1}{n} \sum_{k=0}^{n-1} R_k(\theta)\, \mathrm{d}\mu_c^*(\theta, r), \tag{6.14}$$

where $R_n : \mathcal{A}_c \to \mathbb{R}$ verifies the induction relation $R_{n+1}(\theta) = R_n\big(\theta + R(\theta)\big)$ and $R_1 = R$.

Let $\theta_0 \in \mathcal{A}_c$, $\tilde{\theta}_0$ a lift and $(\tilde{\theta}_k)_{k\in\mathbb{Z}} \in \tilde{\mathfrak{A}}_c$ the associated sequence given by Definition 6.3.17 and $(\theta_k)_{k\in\mathbb{Z}} \in \tilde{\mathcal{A}}_c$ the sequence of projections. For all $k \in \mathbb{Z}$ it then holds that $R(\tilde{\theta}_k) = \tilde{\theta}_{k+1} - \tilde{\theta}_k$. One then readily verifies that $R_n(\theta_0) = R_n(\tilde{\theta}_0) = \tilde{\theta}_{n+1} - \tilde{\theta}_n$ so that $\sum_{k=0}^{n-1} R_k(\theta_0) = \tilde{\theta}_n - \tilde{\theta}_0$. By Theorem 6.1.15 and Corollary 6.3.19

it is inferred that $\frac{1}{n}\sum_{k=0}^{n-1} R_k(\theta)$ uniformly converges to $\rho(c)$ as $n \to +\infty$. It is finally deduced from (4.2.13) that $\rho(\mu_c^*) = \rho(c)$ as desired. □

Remark 6.4.5 The idea behind the previous proof is that all the dynamics of f restricted to the Aubry set $\mathcal{A}_c^*$ can be translated by projecting to a dynamics on $\mathcal{A}_c$. This dynamics is then the restriction of a circle orientation preserving diffeomorphism (by generalizations of Theorem 6.1.15, see [30]). Then, the push forward of any minimizing closed measures on $\mathbb{T}^1$ is invariant by this circle diffeomorphism. Hence the result turns out to be an emanation of Birkhoff's ergodic theorem with this point of view.

Another fact that is apparent from the previous proofs is that if $(\theta, r) \in \mathbb{A}$ is in the support of a minimizing measure μ_c^* as above and if $\tilde{\theta} \in \mathbb{R}$ is a lift of θ, then there exists a unique minimizing sequence $(\tilde{\theta}_k)_{k\in\mathbb{Z}} \in \widetilde{\mathfrak{A}}_c$ such that $\tilde{\theta}_0 = \tilde{\theta}$ and $\tilde{\theta} + r = \tilde{\theta}_1$. Moreover this sequence only depends on the measure μ_c^* (meaning it can be recovered without knowing c).

Now introducing Mather's β function:

Definition 6.4.6 Mather's β function is defined by

$$\forall \rho_0 \in \mathbb{R}, \quad \beta(\rho_0) = -\inf_{\substack{\mu^* \in \mathcal{P}^* \\ \rho(\mu^*) = \rho_0}} \int_{\mathbb{A}} S^*(\theta, r)\, \mathrm{d}\mu^*(\theta, r)$$
$$= \sup_{\substack{\mu^* \in \mathcal{P}^* \\ \rho(\mu^*) = \rho_0}} -\int_{\mathbb{A}} S^*(\theta, r)\, \mathrm{d}\mu^*(\theta, r).$$

Note that for all $\rho_0 \in \mathbb{R}$ the set of closed measures on $\mathbb{A}$ with rotation number ρ_0 is not empty. One may for instance consider the pull back on the circle $\mathbb{T}^1 \times \{\rho_0\}$ of the Lebesgue measure on the circle $\mathbb{T}^1$. Moreover, it is immediate that the rotation number function $\rho : \mathcal{P}^* \to \mathbb{R}$ is linear. Arguing as in the first part of Proposition 6.3.7 immediately yields:

Proposition 6.4.7 *The function β takes values in $\mathbb{R} \cup \{+\infty\}$ and is convex.*

Actually this can be improved by the following Theorem of Mather [165]:

Theorem 6.4.8 *The function β is finite–valued, convex and superlinear. Moreover, α and β are convex dual one another meaning that*

$$\forall \rho_0 \in \mathbb{R}, \quad \beta(\rho_0) = \max_{c\in\mathbb{R}} \rho_0 c - \alpha(c),$$

$$\forall c_0 \in \mathbb{R}, \quad \alpha(c_0) = \max_{\rho\in\mathbb{R}} \rho c_0 - \beta(\rho).$$

Proof We start from the measure characterization of α (given by (6.12) in Proposition 6.4.2) that we rewrite in separating measures according to their rotation number:

$$\alpha(c) = \max_{\mu^* \in \widehat{\mathcal{P}}^*} -\int_{\mathbb{A}} \left[S^*(\theta, r) - cr\right] \mathrm{d}\mu^*(\theta, r)$$

$$= \max_{\varrho \in \mathbb{R}} \sup_{\substack{\mu^* \in \widehat{\mathcal{P}}^* \\ \rho(\mu^*) = \varrho}} c\varrho - \int_{\mathbb{A}} S^*(\theta, r) \,\mathrm{d}\mu^*(\theta, r) = \max_{\varrho \in \mathbb{R}} c\varrho - \beta(\varrho).$$

We recognize here the Fenchel dual of β: $\alpha = \beta^*$ (see [103, 185]). By basic properties of the Fenchel dual, as β^* is everywhere finite, it follows that β is superlinear. Finally, as β is also convex, then $\beta = \beta^{**} = \alpha^*$ as was to be proved. □

We deduce from properties of the Fenchel transform [103, 185], together with Corollary 6.3.19 and Proposition 6.4.4:

Theorem 6.4.9 *The following relations, given c_0 and ρ_0 real numbers, are equivalent:*

- $c_0 \in \partial^- \beta(\rho_0)$;
- $\beta(\rho_0) + \alpha(c_0) = c_0\rho_0$;
- $\rho_0 \in \partial^- \alpha(c_0)$;
- *there exists a minimizing measure $\mu^*_{c_0} \in \mathcal{P}^*$ at cohomology c_0 with rotation number ρ_0,*
- *all minimizing measures $\mu^*_{c_0} \in \mathcal{P}^*$ at cohomology c_0 have rotation number ρ_0,*
- *there exists a minimizing infinite chain, calibrating a weak KAM solution at cohomology c_0 that has rotation number ρ_0,*
- *all minimizing infinite chains, calibrating a weak KAM solution at cohomology c_0 have rotation number ρ_0.*

As for all $c_0 \in \mathbb{R}$, the subdifferential $\partial^-(c_0)$ is a singleton, we deduce that the function α is C^1 and the function β is strictly convex. The function $c \mapsto \rho(c) = \alpha'(c)$ is continuous surjective and non–decreasing.

These last results were first published by Mather in [163] where he attributes them to Aubry.

6.5 Order Properties of Weak KAM Solutions

Results in this section appeared in Arnaud–Zavidovique's works [21, 23]. Some similar statements also can be found in Zhang's work [206] and related results in the setting of the torus $\mathbb{T}^2$ in [79]. Interestingly, non variational versions of some among these results date back to Katznelson and Ornstein [146].

Our first result is that weak KAM solutions and their full pseudographs (Definition 6.3.10) are vertically ordered with respect to the rotation number.

Proposition 6.5.1 *Let $c < c'$ be two cohomology classes such that $\rho(c) < \rho(c')$. Let $u_c : \mathbb{T}^1 \to \mathbb{R}$ be a weak KAM solution at cohomology c and $u_{c'} : \mathbb{T}^1 \to \mathbb{R}$ be a weak KAM solution at cohomology c'. If $\theta \in \mathbb{T}^1$ and r, r' are such that $(\theta, r) \in \mathcal{PG}(c + u_c')$ and $(\theta, r') \in \mathcal{PG}(c' + u_{c'}')$, then $r < r'$.*

In particular, if u_c and $u_{c'}$ are derivable at θ then $c + u_c'(\theta) < c' + u_{c'}'(\theta)$ and the function $\tilde{\theta} \mapsto (\tilde{u}_{c'} - \tilde{u}_c)(\tilde{\theta}) + (c' - c)\tilde{\theta}$ is increasing.

Proof As u_c and $u_{c'}$ are semiconcave, it is enough to prove the following about right and left derivatives: $c' + u_{c'+}'(\theta) > c + u_{c-}'(\theta)$. Set $r_0 = c + u_{c-}'(\theta)$ and $r_0' = c' + u_{c'+}'(\theta)$. Let $\tilde{\theta}_0 = \tilde{\theta}_0' \in \mathbb{R}$ be a lift of θ. For integers $n < 0$, we define $(\tilde{\theta}_n, r_n) = \tilde{f}^n(\tilde{\theta}_0, r_0)$ and $(\tilde{\theta}_n', r_n') = \tilde{f}^n(\tilde{\theta}_0', r_0')$. By Theorem 6.3.8 the sequence $(\tilde{\theta}_n)_{n\leqslant 0}$ (resp. $(\tilde{\theta}_n')_{n\leqslant 0}$) calibrate $\tilde{u}_c$ with cohomology c (resp. $\tilde{u}_{c'}$ with cohomology c'). Hence both sequences are minimizing and by Corollary 6.3.19 verify $\lim\limits_{n\to-\infty} \frac{\tilde{\theta}_n}{n} = \rho(c)$ and $\lim\limits_{n\to-\infty} \frac{\tilde{\theta}_n'}{n} = \rho(c')$. It follows, as $\rho(c) \neq \rho(c')$, that $r_0 \neq r_0'$.

We now argue by contradiction and assume that $r_0' < r_0$. As $\tilde{f}$ twists verticals to the right, $\tilde{f}^{-1}$ twists verticals to the left implying that $\tilde{\theta}_{-1} < \tilde{\theta}_{-1}'$. As $\rho(c') > \rho(c)$ it follows that for large $n < 0$, $\frac{\tilde{\theta}_n'}{n} > \frac{\tilde{\theta}_n}{n}$ and then, for large $n < 0$, $\theta_n' < \theta_n$. We deduce that the sequences $(\tilde{\theta}_n)_{n\leqslant 0}$ and $(\tilde{\theta}_n')_{n\leqslant 0}$ cross at least twice, once at 0 and then once at some negative integer or between two consecutive ones. This contradicts Proposition 6.1.11.

□

With a similar flavor, here is a result on the actions of Lax-Oleinik semigroups (Definition 6.3.5):

Lemma 6.5.2 *Let $c_1 < c_2$ be two real numbers. Let $v_1, v_2 : \mathbb{T}^1 \to \mathbb{R}$ be continuous functions.*

If the function $\theta \mapsto (\tilde{v}_2 - \tilde{v}_1)(\tilde{\theta}) + (c_2 - c_1)\tilde{\theta}$ is non–decreasing, then so is the function $\tilde{\theta} \mapsto (\widetilde{T}^{c_2}\tilde{v}_2 - \widetilde{T}^{c_1}\tilde{v}_1)(\tilde{\theta}) + (c_2 - c_1)\tilde{\theta}$.

Proof Let $\tilde{\theta} < \tilde{\theta}'$ be two real numbers. By definition of the operators T^{c_i} and $\widetilde{T}^{c_i}$ there exist $\tilde{\theta}_2'$ and $\tilde{\theta}_1$ such that

$$\widetilde{T}^{c_2}\tilde{v}_2(\tilde{\theta}') = \tilde{v}_2(\tilde{\theta}_2') + \widetilde{S}(\tilde{\theta}_2', \tilde{\theta}') + c_2(\tilde{\theta}_2' - \tilde{\theta}'),$$

$$\widetilde{T}^{c_1}\tilde{v}_1(\tilde{\theta}) = \tilde{v}_1(\tilde{\theta}_1) + \widetilde{S}(\tilde{\theta}_1, \tilde{\theta}) + c_1(\tilde{\theta}_1 - \tilde{\theta}).$$

There are two cases to consider:

- if $\tilde{\theta}_2' < \tilde{\theta}_1$ we use Aubry & Le Daeron's fundamental Lemma 6.1.10 to obtain

$$\begin{aligned}\widetilde{T}^{c_2}\tilde{v}_2(\tilde{\theta}') + \widetilde{T}^{c_1}\tilde{v}_1(\tilde{\theta}) &= \tilde{v}_2(\tilde{\theta}_2') + \widetilde{S}(\tilde{\theta}_2', \tilde{\theta}') + c_2(\tilde{\theta}_2' - \tilde{\theta}') \\ &\quad + \tilde{v}_1(\tilde{\theta}_1) + \widetilde{S}(\tilde{\theta}_1, \tilde{\theta}) + c_1(\tilde{\theta}_1 - \tilde{\theta}) \\ &> \tilde{v}_2(\tilde{\theta}_2') + \widetilde{S}(\tilde{\theta}_2', \tilde{\theta}) + c_2(\tilde{\theta}_2' - \tilde{\theta}') \\ &\quad + \tilde{v}_1(\tilde{\theta}_1) + \widetilde{S}(\tilde{\theta}_1, \tilde{\theta}') + c_1(\tilde{\theta}_1 - \tilde{\theta}) \\ &\geqslant \widetilde{T}^{c_2}\tilde{v}_2(\tilde{\theta}) + \widetilde{T}^{c_1}\tilde{v}_1(\tilde{\theta}') + (c_2 - c_1)(\tilde{\theta} - \tilde{\theta}').\end{aligned}$$

  After rearranging the terms, this reads

$$\widetilde{T}^{c_2}\tilde{v}_2(\tilde{\theta}') - \widetilde{T}^{c_1}\tilde{v}_1(\tilde{\theta}') + (c_2 - c_1)\tilde{\theta}' > \widetilde{T}^{c_2}\tilde{v}_2(\tilde{\theta}) - \widetilde{T}^{c_1}\tilde{v}_1(\tilde{\theta}) + (c_2 - c_1)\tilde{\theta}.$$

- if $\tilde{\theta}_2' \geqslant \tilde{\theta}_1$ we use the hypothesis on $\tilde{\theta} \mapsto (\tilde{v}_2 - \tilde{v}_1)(\tilde{\theta}) + (c_2 - c_1)\tilde{\theta}$ to show that $\tilde{v}_2(\tilde{\theta}_2') + \tilde{v}_1(\tilde{\theta}_1) \geqslant \tilde{v}_2(\tilde{\theta}_1) + \tilde{v}_1(\tilde{\theta}_2') + (c_2 - c_1)(\tilde{\theta}_1 - \tilde{\theta}_2')$ and then

$$\begin{aligned}\widetilde{T}^{c_2}\tilde{v}_2(\tilde{\theta}') + \widetilde{T}^{c_1}\tilde{v}_1(\tilde{\theta}) &= \tilde{v}_2(\tilde{\theta}_2') + \widetilde{S}(\tilde{\theta}_2', \tilde{\theta}') + c_2(\tilde{\theta}_2' - \tilde{\theta}') \\ &\quad + \tilde{v}_1(\tilde{\theta}_1) + \widetilde{S}(\tilde{\theta}_1, \tilde{\theta}) + c_1(\tilde{\theta}_1 - \tilde{\theta}) \\ &\geqslant \tilde{v}_2(\tilde{\theta}_1) + \widetilde{S}(\tilde{\theta}_2', \tilde{\theta}') + c_2(\tilde{\theta}_1 - \tilde{\theta}') \\ &\quad + \tilde{v}_1(\tilde{\theta}_2') + \widetilde{S}(\tilde{\theta}_1, \tilde{\theta}) + c_1(\tilde{\theta}_2' - \tilde{\theta}) \\ &\geqslant \widetilde{T}^{c_2}\tilde{v}_2(\tilde{\theta}) + \widetilde{T}^{c_1}\tilde{v}_1(\tilde{\theta}') + (c_2 - c_1)(\tilde{\theta} - \tilde{\theta}').\end{aligned}$$

  As before, this gives the result after rearranging terms.

□

We now state our main result of the section. The rest will be devoted to providing elements of its proof. We chose to present the parts which best illustrate discrete weak KAM theory and the first Chapters of this text.

Theorem 6.5.3 *There exists a function $u : \mathbb{T}^1 \times \mathbb{R} \to \mathbb{R}$ that is locally Lipschitz and that verifies the following properties:*

1. *For all $c \in \mathbb{R}$, $u(0, c) = 0$.*
2. *For all $c \in \mathbb{R}$, the function $u_c = u(\cdot, c) : \mathbb{T}^1 \to \mathbb{R}$ is a weak KAM solution at cohomology c.*
3. *If $c < c'$, the function $\tilde{\theta} \mapsto (\tilde{u}_{c'} - \tilde{u}_c)(\tilde{\theta}) + (c' - c)\tilde{\theta}$ is non–decreasing.*

If $u : \mathbb{T}^1 \to \mathbb{R}$ is any continuous function verifying properties 1. and 2. above, then

1. *The map $c \mapsto \mathcal{PG}(c + u_c')$ is continuous for the Hausdorff topology.*

2. *The entire annulus is filled by pseudographs:* $\mathbb{A} = \bigcup_{c\in\mathbb{R}} \mathcal{PG}(c + u_c')$.

The whole proof of this Theorem is quite long and can be found in [21, 23].[5] We give the main steps and ideas bellow.

The first point in the Theorem, $u(0, c) = 0$, is a normalization condition and is easily enforced as the set of weak KAM solutions (at any cohomology class) is invariant by addition of constants.

Let us continue by noticing that a function verifying the first three points of the Theorem is automatically locally Lipschitz. Indeed, weak KAM solutions are locally uniformly in c equiLipschitz in θ (as the costs S^c are). Moreover, if $c \in \mathbb{R}$, then $\tilde{u}_c(1) = \tilde{u}_c(0) = 0$. Then if $\theta \in \mathbb{T}^1$, $0 \leqslant \tilde{\theta} \leqslant 1$ is a lift, and $c < c'$,

$$\begin{aligned} 0 &= (\tilde{u}_{c'} - \tilde{u}_c)(0) + (c' - c) \times 0 \\ &\leqslant (\tilde{u}_{c'} - \tilde{u}_c)(\tilde{\theta}) + (c' - c)\tilde{\theta} \\ &\leqslant (\tilde{u}_{c'} - \tilde{u}_c)(1) + (c' - c) \times 1 = c' - c. \end{aligned}$$

The result follows from

$$c - c' \leqslant -\tilde{\theta}(c' - c) \leqslant (u_{c'} - u_c)(\theta) \leqslant (1 - \tilde{\theta})(c' - c) \leqslant c' - c.$$

Point a. follows from a more general fact, stated in the next Proposition. As we did not find its proof in the literature we provide it. Note that it is valid in any dimension N. In this setting, superdifferentials are linear forms on $\mathbb{R}^N$:

Proposition 6.5.4 *Let $N > 0$ be an integer, $(v_n)_{n\in\mathbb{N}}$ a sequence of (real valued) equi-semiconcave functions defined on $\mathbb{T}^N$. Assume that the sequence $(v_n)_{n\in\mathbb{N}}$ uniformly converges to a function $v : \mathbb{T}^N \to \mathbb{R}$ (that is semiconcave). Then $\big(\mathcal{PG}(v_n)\big)_{n\in\mathbb{N}}$ converges to $\mathcal{PG}(v)$ for the Hausdorff distance.*

Proof By hypothesis there exists a constant $K > 0$ such that all $\tilde{v}_n - K\|\cdot\|^2 : \mathbb{R}^N \to \mathbb{R}$ are strictly concave (where $\|\cdot\|$ denotes the Euclidean norm and $\tilde{v}_n$ the lift of v_n to $\mathbb{R}^N$). It follows that $\tilde{v} - K\|\cdot\|^2 : \mathbb{R}^N \to \mathbb{R}$ is also concave and let us assume it is strictly concave up to taking a slightly larger K. Let $O \subset \mathbb{R}^N$ be a relatively compact, convex, open set. We will show that, restricted to O, $\big(\mathcal{PG}(\tilde{v}_{n|O})\big)_{n\in\mathbb{N}}$ converges to $\mathcal{PG}(u_{|O})$ for the Hausdorff distance, which implies the result.

- Let $X \in O$ and let k_n be an increasing sequence of integers and $(x_{k_n}, p_{k_n}) \in O \times \mathbb{R}^N$ such that $p_{k_n} \in \partial^+\tilde{v}_{k_n}(x_{k_n})$ for all $n \in \mathbb{N}$ and $x_{k_n} \to X$. Assume moreover that $p_{k_n} \to p$ as $n \to +\infty$. We will prove that $p \in \partial^+\tilde{v}(X)$. By hypothesis, for all $n \in \mathbb{N}$, $0 \in \partial^+ w_{k_n}$ where $w_{k_n} : x \mapsto \tilde{v}_{k_n}(x) - K\|x - x_{k_n}\|^2 - p_{k_n}(x)$ is defined on O. It follows that x_{k_n} is a (strict) maximum point of the (strictly)

[5] In those references a result of Mather is used: the β function is derivable on $\mathbb{R} \setminus \mathbb{Q}$. We will give here a strategy of proof that avoids using this result that we will recover later.

concave function w_{k_n}:

$$\forall y \in O, \quad w_{k_n}(y) \leqslant w_{k_n}(x_{k_n}).$$

As $(w_{k_n})_{n\in\mathbb{N}}$ uniformly converges to the function $w : x \mapsto \tilde{v}(x) - K\|x - X\|^2 - p(x)$ we can fix $y \in O$ and pass to the limit in the previous inequality to find that $w(y) \leqslant w(X)$. As this is true for all $y \in O$, X is a maximum point of w which proves, going back to $\tilde{v}$ that $p \in \partial^+\tilde{v}(X)$.

- Let now $X \in O$ and $p \in \partial^+\tilde{v}(X)$. We now define $\varpi : x \mapsto \tilde{v}(x) - K\|x - X\|^2 - p(x)$ that has a strict maximum at X. Let $\varpi_n : x \mapsto \tilde{v}_n(x) - K\|x - X\|^2 - p(x)$ for all integer n, then ϖ_n converges to ϖ uniformly on O. Let $\varepsilon > 0$ small enough such that $B(X, 2\varepsilon) \subset O$. Let $\eta > 0$ such that if $\|x - X\| = \varepsilon$ then $\varpi(x) < \varpi(X) - \eta$. Let n_0 such that for all $n > n_0$, $\|\varpi_n - \varpi\|_{\infty,O} < \eta/3$. If $n > n_0$ and $x \in O$ such that $\|x - X\| = \varepsilon$ it follows that

$$\varpi_n(x) < \varpi(x) + \frac{\eta}{3} < \varpi(X) - \frac{2\eta}{3} < \varpi_n(X) - \frac{\eta}{3}.$$

  It follows that ϖ_n admits a local (hence global) maximum in the ball $B(X, \varepsilon)$ denoted x_n and at which we infer that $0 \in \partial^+\varpi_n(x_n)$.

  Using this argument applied to a decreasing sequence $\varepsilon_k \to 0$ it is easy to construct a sequence x'_n that converges to X defined for $n > n_1$ large enough, and such that $0 \in \partial^+\varpi_n(x'_n)$ for all integer $n > n_1$. Going back to the initial functions, we conclude that

$$\forall n > n_1, \quad p + 2K(x'_n - X) = p_n \in \partial^+\tilde{v}_n(x'_n).$$

  Clearly, $p_n \to p$.

The result follows from the two previous points. □

We now give elements of the construction of a continuous choice of weak KAM solutions with respect to the cohomology class:

Elements of Proof The first fundamental step is: □

Lemma 6.5.5 *Let $\rho_0 \in \mathbb{R}$, we denote by $[a, b] = [\beta'_-(\rho_0), \beta'_+(\rho_0)] = \rho^{-1}(\{\rho_0\})$ (by Theorem 6.4.9). Then there exists a unique weak KAM solution at cohomology a (resp. b), denoted u_a (resp. u_b) such that $u_a(0) = 0$ (resp. $u_b(0) = 0$).*

Let $(c_n)_{n\in\mathbb{N}}$ be a sequence converging to a (resp. converging to b). Let $(v_n)_{n\in\mathbb{N}}$ be a sequence of functions on $\mathbb{T}^1$ such that v_n is a weak KAM solution at cohomology c_n for all n verifying $v_n(0) = 0$. Then the sequence $(v_n)_{n\in\mathbb{N}}$ uniformly converges towards u_a (resp. u_b).

Proof We will prove the result for a, the rest being similar. Let us first consider an increasing sequence $(c_n)_{n\in\mathbb{N}}$ that converges to a. It follows from the hypothesis that $\rho(c_n) < \rho_0$ for all $n \in \mathbb{N}$.

Let $(v_n)_{n\in\mathbb{N}}$ be a sequence of functions on $\mathbb{T}^1$ such that v_n is a weak KAM solution at cohomology c_n for all n verifying $v_n(0) = 0$. The functions v_n are equiLipschitz (by Proposition 2.1.3 and its proof, since the S^{c_n} are) hence by Ascoli's Theorem, the sequence $(v_n)_{n\in\mathbb{N}}$ is relatively compact. As all limit points of $(v_n)_{n\in\mathbb{N}}$ are weak KAM solutions at cohomology a[6] that vanish at 0, it is enough to prove that there is only one such function. To this end, let $u_a : \mathbb{T}^1 \to \mathbb{R}$ be a weak KAM solution at cohomology a such that $u_a(0) = 0$. Let $(k_n)_{n\in\mathbb{N}}$ be an increasing sequence such that $(v_{k_n})_{n\in\mathbb{N}}$ uniformly converges to a function $w : \mathbb{T}^1 \to \mathbb{R}$ that is hence a weak KAM solution at cohomology a such that $w(0) = 0$. Finally, let $\mathcal{D} \subset \mathbb{T}^1$ be a full measure set such that all $(v_n)_{n\in\mathbb{N}}$, u_a, w are derivable on $\mathcal{D}$ and $\widetilde{\mathcal{D}} \subset \mathbb{R}$ its lift. By Proposition 6.5.1, for all $n \in \mathbb{N}$, $s \in \mathcal{D}$, $c_n + v_n'(s) < a + u_a'(s)$. By Proposition 6.5.4, for all $s \in \mathcal{D}$, $w'(s) = \lim_{n\to+\infty} v'_{k_n}(s)$ hence $a + w'(s) \leqslant a + u_a'(s)$. Integrating, it follows that

$$\forall x \in [0, 1], \quad \tilde{u}_a(x) = \int_{\widetilde{\mathcal{D}}\cap[0,x]} \tilde{u}_a'(s)\,\mathrm{d}s \geqslant \int_{\widetilde{\mathcal{D}}\cap[0,x]} \tilde{w}'(s)\,\mathrm{d}s = \tilde{w}(x).$$

For $x = 1$, $\tilde{u}_a(1) = \tilde{w}(1) = 0$. It implies that equality $\tilde{u}_a'(s) = \tilde{w}'(s)$ holds for almost every $s \in [0, 1]$ and then, integrating as above, that $\tilde{u}_a(x) = \tilde{w}(x)$ for all $x \in [0, 1]$. We have thus proved that $u_a = w$ and the uniqueness of u_a follows.

If now $(c_n)_{n\in\mathbb{N}}$ is any sequence converging to a and $(v_n)_{n\in\mathbb{N}}$ is a sequence of weak KAM solutions at cohomology c_n for all n verifying $v_n(0) = 0$ then again, the sequence $(v_n)_{n\in\mathbb{N}}$ is relatively compact and as all its limit points are weak KAM solutions at cohomology a vanishing at 0, by what has been proved, the sequence $(v_n)_{n\in\mathbb{N}}$ converges to u_a.

□

Keeping the notations of the Lemma, a straightforward corollary of the preceding Lemma and Proposition 6.5.1 is,

Corollary 6.5.6 *Let $c \in \mathbb{R}$ and $v_c : \mathbb{T}^1 \to \mathbb{R}$ be a weak KAM solution at cohomology c. Then*

- *$t \mapsto (\tilde{u}_a - \tilde{v}_c)(t) + (a - c)t$ is*
  - *non–increasing if $c > a$,*
  - *increasing if $c < a$.*
- *$t \mapsto (\tilde{u}_b - \tilde{v}_c)(t) + (b - c)t$ is*
  - *non–decreasing if $c < b$,*
  - *decreasing if $c > b$.*

[6] It can be verified with the use of Remark 6.3.2 that $c \mapsto S^c$ is continuous.

We denote by $\mathcal{I} = \{\beta'_-(\rho_0), \beta'_+(\rho_0),\quad \rho_0 \in \mathbb{R}\} \subset \mathbb{R}$ then the previous Lemma tells that any choice of weak KAM solutions u_c at cohomology c for $c \in \mathbb{R}$ such that $u_c(0) = 0$ for all $c \in \mathbb{R}$ is automatically continuous on $\mathcal{I}$. Note that $\mathcal{I}$ is closed as it can alternatively be defined by $\mathcal{I} = \overline{\{\beta'(r),\quad r \in \mathbb{R} \text{ such that } \beta'(r) \text{ exists}\}}$.

The rest of the construction consists in focusing on its complement, $\mathbb{R} \setminus \mathcal{I}$, that is on intervals of the form $\big(\beta'_-(\rho_0), \beta'_+(\rho_0)\big) = (a, b)$ that are not empty (that is where β is not derivable at ρ_0[7]). If this is the case, we use in a crucial way that the Aubry set $\mathcal{A}_c$ does not depend on $c \in \big(\beta'_-(\rho_0), \beta'_+(\rho_0)\big)$ by a result of Massart [157, Proposition 6]. We hence denote it $\mathcal{A}$ until the end of the construction. Let us then interpolate linearly between u_a and u_b restricted to $\mathcal{A}$ and use Proposition 3.2.5 and Theorem 3.2.4. If $t \in (0, 1)$ let $c_t = at + (1-t)b$ and $v_t = tu_a + (1-t)u_b$. As $\alpha(c_t) = t\alpha(a) + (1-t)\alpha(b)$ by duality between α and β, it follows that for all $t \in (0, 1)$, v_t is a critical subsolution for S^{c_t}. Hence by Proposition 3.2.5 and Theorem 3.2.4 there exists a unique weak KAM solution at cohomology c_t, that we note $u_{c_t} = u_t$ for short, such that $u_{t|\mathcal{A}} = v_{t|\mathcal{A}}$. The continuity of $t \mapsto u_t$ is a consequence of uniqueness and Ascoli's Theorem.

Finally, item 3 of the Theorem 6.5.3 follows from successive applications of Lemma 6.5.2 and the alternative characterization: $u_t = \lim\limits_{n\to+\infty} (T^{c_t})^n v_t$.

Again, point a. follows from Proposition 6.5.4. The last thing to prove in the Theorem is b. It uses in a crucial way 2 dimensional topology and Jordan's curve Theorem. It is quite technical and we refer the interested reader to [21, 23]. □

6.6 Structure of Infinite Minimizing Chains

We continue exploring the results of [21, 23] by studying infinite minimizing chains (indexed by non–positive numbers) of the twist map $\tilde{f} : \mathbb{R}^2 \to \mathbb{R}^2$. As we will see, all such minimizing chains calibrate a weak KAM solution, hence fall within the scope of the previous sections. The present results generalize classical Aubry-Mather theory as the latter studies full orbits that are minimizing. After obtaining the results, the authors discovered that many had already appeared in Bangert's [31]. In the latter, Bangert studies Busemann functions that are weak KAM solutions on the universal cover $\mathbb{R}$. We adopt the alternative approach to study and use weak KAM solutions on $\mathbb{T}^1$, as previously defined. Consequently most proofs differ slightly from [31].

We start by a consequence of Theorem 6.5.3 that is a generalization of Corollary 6.3.19.

[7] We will prove later that such ρ_0 are necessarily rational.

Proposition 6.6.1 *Let* $(\tilde{\theta}_k)_{k\leqslant 0} \in \mathbb{R}^{\mathbb{Z}_-}$ *be a minimizing chain. Then there exists* $\rho \in \mathbb{R}$ *such that*

$$\forall k \leqslant 0, \quad |\tilde{\theta}_k - \tilde{\theta}_0 - k\rho| < 2.$$

Proof We denote as usual for all $k \leqslant 0$, $r_k = \partial_2\widetilde{S}(\tilde{\theta}_{k-1}, \tilde{\theta}_k)$ in such a way that $(\tilde{\theta}_k, r_k)_{k\leqslant 0}$ is a piece of orbit of $\tilde{f}$. By Theorem 6.5.3 there exists $c \in \mathbb{R}$ and a weak KAM solution at cohomology c, $u_c : \mathbb{T}^1 \to \mathbb{R}$ such that $(\tilde{\theta}_0, r_0) \in \mathcal{PG}(c + \tilde{u}'_c)$. The result is proved with $\rho = \rho(c)$. Let $r_0^{\pm} = c + \tilde{u}'_{c\pm}(\tilde{\theta}_0)$ (so that $r_0^+ \leqslant r_0^-$). Finally, for $k \geqslant 0$, we set $(\tilde{\theta}_k^{\pm}, r_k^{\pm}) = \tilde{f}^k(\tilde{\theta}_0, r_0^{\pm})$. It follows from Theorem 6.3.8, and the following Remark 6.3.9, that both chains $(\tilde{\theta}_k^+)_{k\leqslant 0}$ and $(\tilde{\theta}_k^-)_{k\leqslant 0}$ calibrate $\tilde{u}_c$ hence are minimizing.

If r_0 coincides with either r_0^+ or r_0^- then the result is a particular case of Corollary 6.3.19. We now assume otherwise which translates to $r_0^+ < r_0 < r_0^-$. As the chains $(\tilde{\theta}_k)_{k\leqslant 0}$ and $(\tilde{\theta}_k^+)_{k\leqslant 0}$ (resp. $(\tilde{\theta}_k)_{k\leqslant 0}$ and $(\tilde{\theta}_k^-)_{k\leqslant 0}$) cross at $k = 0$, they cannot cross anywhere else. We deduce from the twist hypothesis that $\tilde{\theta}_{-1}^- < \tilde{\theta}_{-1} < \tilde{\theta}_{-1}^+$. Hence we conclude from the non–crossing that $\tilde{\theta}_k^- < \tilde{\theta}_k < \tilde{\theta}_k^+$ for all $k < 0$. Finally, from Corollary 6.3.19 we conclude

$$\forall k < 0, \quad k\rho(c) - 2 < \tilde{\theta}_k^- - \tilde{\theta}_0 < \tilde{\theta}_k - \tilde{\theta}_0 < \tilde{\theta}_k^+ - \tilde{\theta}_0 < k\rho(c) + 2.$$

□

Remark 6.6.2 Note that the previous proposition clearly implies that the hypothesis in Proposition 6.1.13, that $|\tilde{\theta}_{i+1} - \tilde{\theta}_i|$ is bounded, is automatically verified by any minimizing chain, hence can be dropped.

We anticipate on the two next sections by stating a result that will be proved later in the text:

Theorem 6.6.3 *Let* $(\tilde{\theta}_k)_{k\leqslant 0} \in \mathbb{R}^{\mathbb{Z}_-}$ *be a minimizing chain. Then there exists a cohomology class* $c \in \mathbb{R}$ *and a weak KAM solution* $u_c : \mathbb{T}^1 \to \mathbb{R}$ *at cohomology* c *such that* $(\tilde{\theta}_k)_{k\leqslant 0}$ *calibrates* $\tilde{u}_c$.

As a consequence, we already state an improvement of Proposition 6.6.1. The proof of the next result uses ideas from Aubry–Mather theory [30] that play a central role in the study of minimizing chains with rational rotation numbers.

Corollary 6.6.4 *Let* $(\tilde{\theta}_k)_{k\leqslant 0} \in \mathbb{R}^{\mathbb{Z}_-}$ *be a minimizing chain. Then there exists* $\rho \in \mathbb{R}$ *such that*

- *for all pairs of integers* (p, q) *with* $q < 0$ *such that* $p/q < \rho$, *then* $\tilde{\theta}_q - \tilde{\theta}_0 < p$,
- *for all pairs of integers* (p, q) *with* $q < 0$ *such that* $p/q > \rho$, *then* $\tilde{\theta}_q - \tilde{\theta}_0 > p$.

In particular,

$$\forall k \leqslant 0, \quad |\tilde{\theta}_k - \tilde{\theta}_0 - k\rho| < 1.$$

Proof Of course, ρ is the same as the one given by Proposition 6.6.1 and is also $\rho(c)$ where c is given by the previous Theorem.

Let us prove the first point, the second is similar. By contradiction, assume the existence of an integer p and $q < 0$ such that $\tilde{\theta}_q - \tilde{\theta}_0 \geqslant p > q\rho$. By the previous Theorem 6.6.3, there exists $c \in \mathbb{R}$ and $u_c : \mathbb{T}^1 \to \mathbb{R}$ such that $(\tilde{\theta}_k)_{k\leqslant 0}$ calibrates $\tilde{u}_c$.

- We first exclude the equality $\tilde{\theta}_q = \tilde{\theta}_0 + p$. Indeed, $\tilde{u}_c$ is derivable at $\tilde{\theta}_q$ by Theorem 6.3.8. Hence assuming by contradiction that $\tilde{\theta}_q = \tilde{\theta}_0 + p$, by periodicity, $\tilde{u}_c$ is also derivable at $\tilde{\theta}_0$ with $\tilde{u}'_c(\tilde{\theta}_0) = \tilde{u}'_c(\tilde{\theta}_q)$. With the usual notations, $r_k = c + \tilde{u}'_c(\tilde{\theta}_k)$, recalling that $(\tilde{\theta}_k, r_k)_{k\leqslant 0}$ is a piece of orbit of $\tilde{f}$, it follows that

$$\forall k \leqslant 0, \quad \tilde{\theta}_{k+q} = \tilde{\theta}_k + p.$$

  By induction, we deduce that $\tilde{\theta}_{nq} = \tilde{\theta}_0 + np$ for all $n > 0$ and finally, dividing by nq and letting $n \to +\infty$ we conclude that $\rho = \frac{p}{q}$. This contradicts the right inequality $p > q\rho$.
- We are left with the hypothesis that $\tilde{\theta}_q - \tilde{\theta}_0 > p > q\rho$. By periodicity, the chain $(\tilde{\theta}_{q+k} - p)_{k\leqslant 0}$ also calibrates $\tilde{u}_c$. By Lemma 6.3.12 we deduce that $\tilde{\theta}_{k+q} - p > \tilde{\theta}_k$ for all $k \leqslant 0$. By induction we readily obtain that for all $k \leqslant 0$, the sequence $(\tilde{\theta}_{k+nq} - np)_{n\geqslant 0}$ is increasing. Applying for $k = 0$ and dividing by $nq < 0$ yields $\dfrac{\tilde{\theta}_{nq}}{nq} - \dfrac{p}{q} < \dfrac{\tilde{\theta}_0}{nq}$. Finally, letting $n \to +\infty$ entails $\rho \leqslant \dfrac{p}{q}$ that is again a contradiction.

To prove the final statement, let us argue by contradiction and assume that there exists $q < 0$ such that $|\tilde{\theta}_q - \tilde{\theta}_0 - q\rho| \geqslant 1$. Then one of the following holds:

$$\exists p \in \mathbb{Z}, \quad \tilde{\theta}_q - \tilde{\theta}_0 \leqslant p < q\rho,$$

$$\exists p \in \mathbb{Z}, \quad \tilde{\theta}_q - \tilde{\theta}_0 \geqslant p > q\rho,$$

both impossible by what was proved above. □

We now turn to making Proposition 6.1.13 more precise.

Proposition 6.6.5 *Let $c \in \mathbb{R}$ be a cohomology class. Let $(\tilde{\theta}_i)_{i\leqslant 0}$ and $(\tilde{\theta}^c_i)_{i\leqslant 0}$ be two minimizing chains such that*

- $\tilde{\theta}_0 = \tilde{\theta}^c_0$,
- $(\tilde{\theta}_i)_{i\leqslant 0}$ *and* $(\tilde{\theta}^c_i)_{i\leqslant 0}$ *are* α*-asymptotic,*
- *there exists a weak KAM solution $u_c : \mathbb{T}^1 \to \mathbb{R}$ at cohomology c, such that $(\tilde{\theta}^c_i)_{i\leqslant 0}$ calibrates $\tilde{u}_c$.*

Then $(\tilde{\theta}_i)_{i\leqslant 0}$ also calibrates $\tilde{u}_c$.

Proof We argue by contradiction assuming that there exist $n_0 < 0$ and $\varepsilon > 0$ such that

$$\tilde{u}_c(\tilde{\theta}_0) < \tilde{u}_c(\tilde{\theta}_{n_0}) + \sum_{i=n_0}^{-1} \widetilde{S}(\tilde{\theta}_i, \tilde{\theta}_{i+1}) + c(\tilde{\theta}_{n_0} - \tilde{\theta}_0) + |n_0|\alpha(c) - \varepsilon.$$

As for $n < n_0$,

$$\tilde{u}_c(\tilde{\theta}_{n_0}) \leqslant \tilde{u}_c(\tilde{\theta}_n) + \sum_{i=n}^{n_0-1} \widetilde{S}(\tilde{\theta}_i, \tilde{\theta}_{i+1}) + c(\tilde{\theta}_n - \tilde{\theta}_{n_0}) + |n_0 - n|\alpha(c),$$

by summing the two previous inequalities, we deduce that

$$\forall n \leqslant n_0, \quad \tilde{u}_c(\tilde{\theta}_0) < \tilde{u}_c(\tilde{\theta}_n) + \sum_{i=n}^{-1} \widetilde{S}(\tilde{\theta}_i, \tilde{\theta}_{i+1}) + c(\tilde{\theta}_n - \tilde{\theta}_0) + |n|\alpha(c) - \varepsilon.$$

By uniform continuity of $\widetilde{S}$ on compact sets and Remark 6.6.2 there exists $n_1 < n_0$ such that

$$\forall i < n_1, \quad |\widetilde{S}(\tilde{\theta}_i, \tilde{\theta}_{i+1}) - \widetilde{S}(\tilde{\theta}_i^c, \tilde{\theta}_{i+1})| < \frac{\varepsilon}{3},$$

and $|\tilde{u}_c(\tilde{\theta}_n) - \tilde{u}_c(\tilde{\theta}_n^c) + c(\tilde{\theta}_n - \tilde{\theta}_n^c)| < \varepsilon/3$. It follows that for $n < n_1$,

$$\begin{aligned}
\widetilde{S}(\tilde{\theta}_n, \tilde{\theta}_{n+1}^c) + \sum_{i=n_0}^{-1} \widetilde{S}(\tilde{\theta}_i^c, \tilde{\theta}_{i+1}^c) &\leqslant \sum_{i=n}^{-1} \widetilde{S}(\tilde{\theta}_i^c, \tilde{\theta}_{i+1}^c) + \frac{\varepsilon}{3} \\
&= \tilde{u}_c(\tilde{\theta}_0^c) - \tilde{u}_c(\tilde{\theta}_n^c) + c(\tilde{\theta}_0^c - \tilde{\theta}_n^c) - |n|\alpha(c) + \frac{\varepsilon}{3} \\
&\leqslant \tilde{u}_c(\tilde{\theta}_0) - \tilde{u}_c(\tilde{\theta}_n) + c(\tilde{\theta}_0 - \tilde{\theta}_n) - |n|\alpha(c) + \frac{2\varepsilon}{3} \\
&\leqslant \sum_{i=n}^{-1} \widetilde{S}(\tilde{\theta}_i, \tilde{\theta}_{i+1}) - \frac{\varepsilon}{3}.
\end{aligned}$$

This contradicts the fact that $(\tilde{\theta}_i)_{i\in[n,0]}$ is minimizing.

□

We are now ready to show strong results on infinite minimizing chains according to their rotation number. As we will see, the nature of those results depends strongly on its rationality. The next two sections, are devoted to recalling facts about homeomorphisms of the circle and about Aubry sets of twist maps. Many of those results are enunciated and proved in [30] as well as in many surveys about Poincaré and Denjoy theories [195].

6.6.1 Irrational Rotation Number

We assume in this paragraph that $\rho_0 \in \mathbb{R} \setminus \mathbb{Q}$. Let $c_0 \in \mathbb{R}$ such that $\rho(c_0) = \rho_0$, which exists thanks to Theorems 6.4.8 and 6.4.9. We recall without proof [195]:

Theorem 6.6.6 *Let $\tilde{g} : \mathbb{R} \to \mathbb{R}$ be the lift of an orientation preserving circle homeomorphism $g : \mathbb{T}^1 \to \mathbb{T}^1$ of rotation number $\rho_0 \in \mathbb{R} \setminus \mathbb{Q}$. Then there exists a (unique up to an integer) non–decreasing $\tilde{\varphi} \to \mathbb{R} \to \mathbb{R}$ such that*

$$\forall x \in \mathbb{R}, \quad \tilde{\varphi}(x+1) = \tilde{\varphi}(x) + 1,$$

that semi–conjugates $\tilde{g}$ to the translation by ρ_0:

$$\forall x \in \mathbb{R}, \quad \tilde{\varphi} \circ \tilde{g}(x) = \varphi(x) + \rho_0.$$

There is then an alternative:

1. *Either $\tilde{\varphi}$ is increasing in which case $\tilde{g}$ is conjugated to an irrational rotation and all its orbits are recurrent.*
2. *Either $\tilde{\varphi}$ is not increasing, in which case the set of recurrent orbits g is a Cantor set denoted by $\mathcal{K} \subset \mathbb{T}^1$ that lifts to $\widetilde{\mathcal{K}} \subset \mathbb{R}$. If $(\tilde{a}, \tilde{b})$ is a connected component of $\mathbb{R} \setminus \widetilde{\mathcal{K}}$ and $(a, b) \in \mathbb{T}^1$ its projection, the images $g^k(a, b)$ are mutually disjoint as $k \in \mathbb{Z}$. In particular,*

$$\sum_{k \in \mathbb{Z}} \left(\tilde{g}^k(\tilde{b}) - \tilde{g}^k(\tilde{a})\right) \leqslant 1.$$

In all cases there is a unique g–invariant probability measure that has full support in the first case and support $\mathcal{K}$ in the second case. Finally, if $x_0 \in \mathbb{T}^1$ then the α–limit set of the orbit of x_0 is $\mathbb{T}^1$ in the first case and $\mathcal{K}$ in the second one.

We then recall related results on twist maps, the proof of which are omitted, but can be found in [30, 163, 172] and a glimpse of which appears in Remark 6.4.5:

Theorem 6.6.7 *Let $\mu^*_{\rho_0}$ be a Mather minimizing measure verifying* (6.12) *for c_0. Then by Theorem 6.4.9, $\mu^*_{\rho_0}$ is minimizing for all $c \in \rho^{-1}(\{\rho_0\})$. In particular its support verifies*

$$\forall c \in \rho^{-1}(\{\rho_0\}), \quad \pi_1\left(\mathrm{supp}(\mu^*_{\rho_0})\right) \subset \mathcal{A}_c.$$

Let $\theta_0^{\rho_0} \in \text{supp}(\mu^*_{\rho_0})$ *and let* $\tilde{\theta}_0^{\rho_0} \in \mathbb{R}$ *a lift and* $(\tilde{\theta}_k^{\rho_0})_{k\in\mathbb{Z}} \in \widetilde{\mathfrak{A}}_{c_0}$ *the associated minimizing sequence. As seen in the proof of Proposition 6.4.4 this sequence only depends on* $\mu^*_{\rho_0}$ *hence* $(\tilde{\theta}_k^{\rho_0})_{k\in\mathbb{Z}} \in \widetilde{\mathfrak{A}}_c$ *for all* $c \in \rho^{-1}(\{\rho_0\})$.

Let $\tilde{g} : \mathbb{R} \to \mathbb{R}$ *the associated map given by Theorem 6.1.15 and* $g : \mathbb{T}^1 \to \mathbb{T}^1$ *its projection.*[8] *Then* $\pi_{1*}\mu^*_{\rho_0}$ *is* g*–invariant. In particular* $\pi_1\big(\text{supp}(\mu^*_{\rho_0})\big)$ *is either* $\mathbb{T}^1$ *or the* g*–invariant Cantor set* $\mathcal{K}$*. Moreover,* $\tilde{g}$ *only depends on* $\mu^*_{\rho_0}$.

The main result of this section is then

Theorem 6.6.8 *Let* $(\tilde{\theta}_k)_{k\leqslant 0}$ *be a minimizing chain of rotation number* $\rho_0 \in \mathbb{R} \setminus \mathbb{Q}$*. Let* $u_{c_0} : \mathbb{T}^1 \to \mathbb{R}$ *be a weak KAM solution at cohomology* $c_0 \in \rho^{-1}(\{\rho_0\})$*. Then* $(\tilde{\theta}_k)_{k\leqslant 0}$ *calibrates* $\tilde{u}_{c_0}$.

Proof As always, $r_k = \partial_2\widetilde{S}(\tilde{\theta}_{k-1}, \tilde{\theta}_k)$. Arguing as in the proof of Proposition 6.6.1, there exists $c \in \mathbb{R}$ and $u_c : \mathbb{T}^1 \to \mathbb{R}$ a weak KAM solution at cohomology c such that $(\tilde{\theta}_0, r_0) \in \mathcal{PG}(c + \tilde{u}'_c)$ and $\rho_0 = \rho(c)$. We adopt the same notations as in Proposition 6.6.1 setting $r_0^{\pm} = c + \tilde{u}'_{c\pm}(\tilde{\theta}_0)$ (so that $r_0^+ \leqslant r_0^-$) and for $k \leqslant 0$, $(\tilde{\theta}_k^{\pm}, r_k^{\pm}) = \tilde{f}^k(\tilde{\theta}_0, r_0^{\pm})$.

Finally, let $y_0^- \in \pi^{-1} \circ \pi_1\big(\text{supp}(\mu^*_{\rho_0})\big) \subset \mathbb{R}$ (resp. y_0^+) be the biggest element of $\pi^{-1} \circ \pi_1\big(\text{supp}(\mu^*_{\rho_0})\big)$ such that $y_0^- \leqslant \tilde{\theta}_0$ (resp. smallest element of $\pi^{-1} \circ \pi_1\big(\text{supp}(\mu^*_{\rho_0})\big)$ such that $y_0^+ \geqslant \tilde{\theta}_0\big)$. Let then $(y_k^-)_{k\in\mathbb{Z}} \in \widetilde{\mathfrak{A}}_c$ and $(y_k^+)_{k\in\mathbb{Z}} \in \widetilde{\mathfrak{A}}_c$ the associated minimizing sequences that calibrate $\tilde{u}_c$ (by Theorem 6.6.7). As in the proof of Proposition 6.6.1, for all $k \leqslant 0$,

$$y_k^- \leqslant \tilde{\theta}_k^- \leqslant \tilde{\theta}_k \leqslant \tilde{\theta}_k^+ \leqslant y_k^+.$$

By Theorem 6.6.7, $y_k^- - y_k^+ \to 0$ as $k \to -\infty$ hence all the present sequences are α–asymptotic.

Let now $(\tilde{\theta}_k^{c_0})_{k\leqslant 0}$ be a calibrating chain for $\tilde{u}_{c_0}$ with $\tilde{\theta}_0^{c_0} = \tilde{\theta}_0$ (that exists thanks to Theorem 6.3.8). As $(y_k^{\pm})_{k\in\mathbb{Z}}$ both calibrate $\tilde{u}_{c_0}$ the same reasoning as above yields that $(\tilde{\theta}_k^{c_0})_{k\leqslant 0}$ and $(y_k^{\pm})_{k\in\mathbb{Z}}$ are all α–asymptotic. Finally, Proposition 6.6.5 applies and $(\tilde{\theta}_k)_{k\leqslant 0}$ does indeed calibrate $\tilde{u}_{c_0}$. □

As a Corollary, we obtain uniqueness of weak KAM solutions up to constants with irrational rotation number, and a Theorem due to Mather [164] and Bangert [31]. A different proof also can be found in [41].

Corollary 6.6.9 *If* $\rho_0 \in \mathbb{R} \setminus \mathbb{Q}$ *then* $\rho^{-1}(\{\rho_0\}) = \{c_0\}$ *is a singleton, meaning that* β *is derivable at* ρ_0*. Moreover, if* $u : \mathbb{T}^1 \to \mathbb{R}$ *and* $v : \mathbb{T}^1 \to \mathbb{R}$ *are weak KAM solutions at cohomology* c_0 *then* $u - v$ *is constant.*

[8] The results of Mather proven in [30] actually show that the maps $\tilde{g}$ and g can be chosen only depending on ρ_0 and that the measure $\mu^*_{\rho_0}$ is unique.

Proof We will prove both statements at once. Let $\{c_0, c_1\} \in \rho^{-1}(\{\rho_0\})$ and let $u_0 : \mathbb{T}^1 \to \mathbb{R}$ be a weak KAM solution at cohomology c_0 and $u_1 : \mathbb{T}^1 \to \mathbb{R}$ be a weak KAM solution at cohomology c_1. Let $D \subset \mathbb{T}^1$ be the set of points where both u_0 and u_1 are derivable. It is of full Lebesgue measure as its complement is countable. Let $\theta_0 \in D$ and $\tilde{\theta}_0$ a lift. By Theorem 6.3.8, there exist unique chains $(\tilde{\theta}_k^i)_{k\leqslant 0}$ for $i \in \{0, 1\}$ calibrating respectively $\tilde{u}_0$ and $\tilde{u}_1$ and such that $\tilde{\theta}_0^i = \tilde{\theta}_0$ for $i \in \{0, 1\}$.

Moreover, $c_i + u_i'(\theta_0) = \partial_2\widetilde{S}(\tilde{\theta}_{-1}^i, \tilde{\theta}_0^i)$. Applying twice Theorem 6.6.8 we discover that $(\tilde{\theta}_k^0)_{k\leqslant 0}$ calibrates $\tilde{u}_1$ and $(\tilde{\theta}_k^1)_{k\leqslant 0}$ calibrates $\tilde{u}_0$. Hence by uniqueness, $(\tilde{\theta}_k^0)_{k\leqslant 0} = (\tilde{\theta}_k^1)_{k\leqslant 0}$ and $c_0 + u_0'(\theta_0) = c_1 + u_1'(\theta_0)$.

Integrating on $\mathbb{T}^1$ it follows that

$$c_0 = \int_{\mathbb{T}^1} \big(c_0 + u_0'(s)\big)\mathrm{d}s = \int_{\mathbb{T}^1} \big(c_1 + u_1'(s)\big)\mathrm{d}s = c_1.$$

As a conclusion, since u_0 and u_1 are Lipschitz with almost everywhere equal derivatives, $u_1 - u_0$ is a constant function. □

As there exists a weak KAM solution u_{c_0} associated to the irrational rotation number ρ_0, we know that for every $\tilde{\theta}_0 \in \mathbb{R}$ there exists a minimizing chain $(\tilde{\theta}_k)_{k\leqslant 0}$ starting at $\tilde{\theta}_0$ with rotation number ρ_0 (one calibrating u_{c_0}). This next and last corollary on the contrary states that there are not too many such minimizing chains:

Corollary 6.6.10 *Let $\rho_0 \in \mathbb{R}\setminus\mathbb{Q}$. For all $\tilde{\theta} \in \mathbb{R}$ there exists at most one minimizing chain $(\tilde{\theta}_k)_{k\leqslant 0}$ with rotation number ρ_0 such that $(\tilde{\theta}_0, r_0) \in \tilde{f}(V_{\tilde{\theta}})$ where $(\tilde{\theta}_k, r_k)_{k\leqslant 0}$ is the backward $\tilde{f}$–orbit associated to $(\tilde{\theta}_k)_{k\leqslant 0}$ and $V_{\tilde{\theta}} = \{(\tilde{\theta}, r),\ r \in \mathbb{R}\} \subset \mathbb{R}\times\mathbb{R}$ is the vertical above $\tilde{\theta}$.*

Proof Using the notations of the Corollary, it was just established that, for such a minimizing chain, $(\tilde{\theta}_0, r_0) \in \mathcal{PG}(c_0 + \tilde{u}_{c_0}')$ where $\rho^{-1}(\{\rho_0\}) = \{c_0\}$ and $\mathcal{PG}(c_0 + \tilde{u}_{c_0}')$ is the unique pseudograph associated to c_0.

By Corollary 6.3.13, $f^{-1}\big(\mathcal{PG}(c_0+\tilde{u}_{c_0}')\big)$ is a graph, meaning that $\tilde{f}^{-1}\big(\mathcal{PG}(c_0+\tilde{u}_{c_0}')\big) \cap V_{\tilde{\theta}}$ is a singleton and consequently, so is $\mathcal{PG}(c_0 + \tilde{u}_{c_0}') \cap \tilde{f}(V_{\tilde{\theta}})$. The result follows. □

6.6.2 Rational Rotation Number

We assume in this paragraph that $\rho_0 = \frac{p}{q}$ is rational, written in irreducible form with $q < 0$. We denote by $[a, b] = \rho^{-1}(\{\rho_0\})$. Let us first recall some classical results from Aubry–Mather theory [30]:

Theorem 6.6.11 *Let $\mu_{\rho_0}^*$ be a Mather minimizing measure verifying* (6.12) *for some $c_0 \in [a, b]$. Then by Theorem 6.4.9, $\mu_{\rho_0}^*$ is minimizing for all $c \in \rho^{-1}(\{\rho_0\})$. In* *particular its support verifies*

$$\forall c \in \rho^{-1}(\{\rho_0\}), \quad \pi_1\big(\mathrm{supp}(\mu^*_{\rho_0})\big) \subset \mathcal{A}_c.$$

Let $\theta_0^{\rho_0} \in \mathrm{supp}(\mu^*_{\rho_0})$ *and let* $\tilde\theta_0^{\rho_0} \in \mathbb{R}$ *a lift and* $(\tilde\theta_k^{\rho_0})_{k\in\mathbb{Z}} \in \widetilde{\mathfrak{A}}_{c_0}$ *the associated minimizing sequence. As seen in the proof of Proposition 6.4.4 this sequence only depends on* $\mu^*_{\rho_0}$ *hence* $(\tilde\theta_k^{\rho_0})_{k\in\mathbb{Z}} \in \widetilde{\mathfrak{A}}_c$ *for all* $c \in \rho^{-1}(\{\rho_0\})$.

It verifies

$$\forall k \in \mathbb{Z}, \quad \tilde\theta^{\rho_0}_{k+q} = \tilde\theta^{\rho_0}_k + p, \tag{6.15}$$

and hence projects to a q*–periodic sequence on* $\mathbb{T}^1$.

Reciprocally, if a minimizing sequence $(\tilde\theta_k)_{k\in\mathbb{Z}} \in \mathbb{R}^{\mathbb{Z}}$ *satisfies* (6.15)*, then it is in any Aubry set* $\widetilde{\mathfrak{A}}_c$ *for* $c \in [a, b]$ *and the measure*

$$\mu^* = \frac{1}{q}\sum_{k=0}^{q-1} \delta_{(\theta_k, \tilde\theta_{k+1} - \tilde\theta_k)},$$

verifies that μ^* *is minimizing (satisfies* (6.12)*) for all* $c \in [a, b]$.

A notion that stems from the previous Theorem is:

Definition 6.6.12 A sequence $(\tilde\theta_k)_{k\in\mathbb{Z}} \in \mathbb{R}^{\mathbb{Z}}$ that verifies

$$\forall k \in \mathbb{Z}, \quad \tilde\theta_{k+q} = \tilde\theta_k + p,$$

is said to be of type (p, q).

It follows from the previous Theorem that it makes sense to denote the projected Mather set $\mathcal{M}_{\rho_0}$ as $\mathcal{M}_c$ does not depend on the choice of $c \in \rho^{-1}(\{\rho_0\})$. And $\mathfrak{M}_{\rho_0}$ denotes its lift to $\mathbb{R}$ as well.

We recall that by Lemma 6.5.5, there exists a unique weak KAM solution at cohomology a (resp. b) that vanishes at $0 \in \mathbb{T}^1$, and that we denote $u_a : \mathbb{T}^1 \to \mathbb{R}$ (resp. $u_b : \mathbb{T}^1 \to \mathbb{R}$). The main result of this section can be stated as follows:

Theorem 6.6.13 *Let* $(\tilde\theta_k)_{k\leqslant 0}$ *be a minimizing chain with rotation number* $\rho_0 = \frac{p}{q}$. *Assume that* $(\tilde\theta_k)_{k\leqslant 0}$ *is not of type* (p, q)*. Then* $\tilde\theta_0 \neq \tilde\theta_q - p$ *and one of the following assertions holds:*

1. $\tilde\theta_0 < \tilde\theta_q - p$ *and* $(\tilde\theta_k)_{k\leqslant 0}$ *calibrates* $\tilde u_a$,
2. $\tilde\theta_0 > \tilde\theta_q - p$ *and* $(\tilde\theta_k)_{k\leqslant 0}$ *calibrates* $\tilde u_b$.

The proof of this Theorem will be done in several steps and provides more precise details about the behavior of minimizing chains and their links with weak KAM solutions.

We start by the following non–crossing lemma for minimizing chains with rotation number $\rho_0 = \frac{p}{q}$. It is reminiscent of Corollary 6.6.4:

Proposition 6.6.14 *Let $(\tilde{\theta}_k)_{k\leqslant 0}$ be a minimizing chain with rotation number $\rho_0 = \frac{p}{q}$. Assume that $(\tilde{\theta}_k)_{k\leqslant 0}$ is not of type (p,q). Then the minimizing chains $(\tilde{\theta}_k)_{k\leqslant 0}$ and $(\tilde{\theta}_{k+q} - p)_{k\leqslant 0}$ do not cross.*

Moreover,

- *if $\tilde{\theta}_0 < \tilde{\theta}_q - p$ then $(\tilde{\theta}_k)_{k\leqslant 0}$ calibrates $\tilde{u}_a$,*
- *if $\tilde{\theta}_0 > \tilde{\theta}_q - p$ then $(\tilde{\theta}_k)_{k\leqslant 0}$ calibrates $\tilde{u}_b$.*

Proof As $(\tilde{\theta}_k)_{k\leqslant 0}$ and $(\tilde{\theta}_{k+q} - p)_{k\leqslant 0}$ cross at most once, there exists $n_0 \leqslant 0$ such that one of the following holds

$$\forall k < n_0, \quad \tilde{\theta}_k < \tilde{\theta}_{k+q} - p,$$

$$\forall k < n_0, \quad \tilde{\theta}_k > \tilde{\theta}_{k+q} - p.$$

Let us deal with the first case, the second being similar. By induction, it follows that for all $k < n_0$, the sequence $(\tilde{\theta}_{k+mq} - mp)_{m\geqslant 0}$ is increasing.

Let us introduce some notations. Once more, for all $k \leqslant 0$, we set $r_k = \partial_2\tilde{S}(\tilde{\theta}_{k-1}, \tilde{\theta}_k)$ in such a way that $(\tilde{\theta}_k, r_k)_{k\leqslant 0}$ is a piece of orbit of $\tilde{f}$. Let $c \in \mathbb{R}$ such that $(\tilde{\theta}_0, r_0) \in \mathcal{PG}(c + \tilde{u}_c')$ for some weak KAM solution $u_c : \mathbb{T}^1 \to \mathbb{R}$ at cohomology c. Then as in Proposition 6.6.1, $\rho(c) = \frac{p}{q}$ and $c \in [a,b]$. Moreover, $\tilde{\theta}_0 \notin \mathfrak{M}_{\rho_0}$. Indeed, if it were the case, as $\tilde{u}_c$ is derivable on $\mathfrak{M}_{\rho_0}$, we would have $(\tilde{\theta}_k, r_k)_{k\leqslant 0} = \big(\tilde{f}^k(\tilde{\theta}_0, \tilde{u}_c'(\tilde{\theta}_0) + c)\big)_{k\leqslant 0}$ that is of type (p,q) by Theorem 6.6.11.

It is then denoted $y_0^- = \max\{y \in \mathfrak{M}_{\rho_0},\ y < \tilde{\theta}_0\}$ and $y_0^+ = \min\{y \in \mathfrak{M}_{\rho_0},\ y > \tilde{\theta}_0\}$. Finally, let $(y_k^-)_{k\in\mathbb{Z}}$ and $(y_k^+)_{k\in\mathbb{Z}}$ the associated minimizing sequences, that are of type (p,q). By Theorem 6.6.11 the sequences $(y_k^\pm)_{k\in\mathbb{Z}}$ calibrate $\tilde{u}_c$. First note that $y_0^- < \tilde{\theta}_0 < y_0^+$ because $\mathfrak{M}_{\rho_0}$ is closed. Moreover, arguing as in Proposition 6.6.1 yields that

$$\forall k \leqslant 0, \quad y_k^- < \tilde{\theta}_k < y_k^+.$$

As

$$\forall m \geqslant 0, \quad y_k^- = y_{k+mq}^- - mp < \tilde{\theta}_{k+mq} - mp < y_{k+mq}^+ - mp = y_k^+, \tag{6.16}$$

we deduce that for all $k \in \mathbb{Z}$, the sequence $(\tilde{\theta}_{k+mq} - mp)_{m\geqslant -k/q}$ is bounded, increasing for m large enough, hence it converges to some y_k. Moreover, the sequence $(y_k)_{k\in\mathbb{Z}}$ is minimizing as a limit of minimizing chains and verifies

$$\forall k \in \mathbb{Z}, \quad y_{k+q} - p = \lim_{m\to+\infty} \tilde{\theta}_{k+mq+q} - mp - p = \lim_{m\to+\infty} \tilde{\theta}_{k+(m+1)q} - (m+1)p = y_k.$$

It follows from Theorem 6.6.11 that $y_k \in \mathfrak{M}_{\rho_0}$ and that $(y_k)_{k\in\mathbb{Z}}$ calibrates $\tilde{u}_c$. By (6.16), $\tilde{\theta}_k < y_k \leqslant y_k^+$. If $y_k < y_k^+$ it follows from Lemma 6.3.12 that $\tilde{\theta}_0 < y_0 < y_0^+$ thus contradicting the definition of y_0^+. From (6.16) comes that $(\tilde{\theta}_k)_{k\leqslant 0}$, $(\tilde{\theta}_{k+q} - p)_{k\leqslant 0}$ and $(y_k^+)_{k\in\mathbb{Z}}$ are α–asymptotic. It is deduced from Proposition 6.1.11 that $(\tilde{\theta}_k)_{k\leqslant 0}$ and $(\tilde{\theta}_{k+q} - p)_{k\leqslant 0}$ cannot cross (as $(\tilde{\theta}_{k+q} - p)_{k\leqslant 0}$ is a strict subchain of a minimizing chain).

To finish the proof, let $(\tilde{\theta}_k^a)_{k\leqslant 0}$ be the minimizing chain calibrating $\tilde{u}_a$, with $\tilde{\theta}_0^a = \tilde{\theta}_0$ and $\partial_2\widetilde{S}(\tilde{\theta}_{-1}^a, \tilde{\theta}_0^a) = a + \tilde{u}'_{a+}(\tilde{\theta}_0)$. As $r_0 - c \in \partial^+\tilde{u}_c(\tilde{\theta}_0)$, it follows from Corollary 6.5.6 that $r_0 \geqslant a + \tilde{u}'_{a+}(\tilde{\theta}_0)$. Either there is equality, in which case $(\tilde{\theta}_k^a)_{k\leqslant 0} = (\tilde{\theta}_k)_{k\leqslant 0}$ calibrates $\tilde{u}_a$, or the inequality is strict and the twist condition entails that $\tilde{\theta}_{-1}^a > \tilde{\theta}_{-1}$. As the two minimizing chains do not cross anymore, and using that $(y_k^-)_{k\in\mathbb{Z}}$ calibrates $\tilde{u}_a$, it follows that

$$\forall k \leqslant 0, \quad \tilde{\theta}_k < \tilde{\theta}_k^a < y_k^+.$$

Hence $(\tilde{\theta}_k^a)_{k\leqslant 0}$ and $(\tilde{\theta}_k)_{k\leqslant 0}$ are α–asymptotic and by Proposition 6.6.5, $(\tilde{\theta}_k)_{k\leqslant 0}$ calibrates $\tilde{u}_a$. □

A Corollary of this proof is:

Corollary 6.6.15 *Let $(\tilde{\theta}_k)_{k\leqslant 0}$ be a minimizing chain with rotation number $\rho_0 = \frac{p}{q}$. Assume that $(\tilde{\theta}_k)_{k\leqslant 0}$ is not of type (p, q) (Definition 6.6.12). Let $(y_k^\pm)_{k\in\mathbb{Z}}$ be the closest orbits of $\mathfrak{M}_{\rho_0}$ such that $y_k^- < \tilde{\theta}_k < y_k^+$ for $k \leqslant 0$ as in the proof of Proposition 6.6.14. Then*

- *either $\tilde{\theta}_0 < \tilde{\theta}_q - p$, $(\tilde{\theta}_k)_{k\leqslant 0}$ and $(y_k^+)_{k\leqslant 0}$ are α–asymptotic,*
- *either $\tilde{\theta}_0 > \tilde{\theta}_q - p$, $(\tilde{\theta}_k)_{k\leqslant 0}$ and $(y_k^-)_{k\leqslant 0}$ are α–asymptotic.*

As another consequence, we derive that minimizing chains with rotation number p/q are quite rare, same as was established for irrational rotation numbers in Corollary 6.6.10:

Theorem 6.6.16 *Let $\widetilde{\Theta} \in \mathbb{R}$ and $V_{\widetilde{\Theta}} = \{\widetilde{\Theta}\} \times \mathbb{R} \subset \mathbb{R} \times \mathbb{R}$. Then there are at most two minimizing half orbits of $\tilde{f}$, $(\tilde{\theta}_k, r_k)_{k\leqslant 0}$ such that $(\tilde{\theta}_0, r_0) \in \tilde{f}(V_{\widetilde{\Theta}})$ and any corresponding minimizing chain $(\tilde{\theta}_k)_{k\leqslant 0}$ has rotation number $\rho_0 = \frac{p}{q}$.*

Proof It was just settled in Proposition 6.6.14 that the chain $(\tilde{\theta}_k)_{k\leqslant 0}$ calibrates $\tilde{u}_a$ or $\tilde{u}_b$. Moreover, by Theorem 6.3.8, $(\tilde{\theta}_0, r_0) \in \mathcal{PG}(a + \tilde{u}'_a) \cup \mathcal{PG}(b + \tilde{u}'_b)$. Finally, $\tilde{f}^{-1}(\tilde{\theta}_0, r_0) \in \tilde{f}^{-1}\big(\mathcal{PG}(a + \tilde{u}'_a)\big) \cup \tilde{f}^{-1}\big(\mathcal{PG}(b + \tilde{u}'_b)\big)$. By Corollary 6.3.13, both $f^{-1}\big(\mathcal{PG}(a + \tilde{u}'_a)\big)$ and $\tilde{f}^{-1}\big(\mathcal{PG}(b + \tilde{u}'_b)\big)$ are graphs, hence $f^{-1}\big(\mathcal{PG}(a + \tilde{u}'_a)\big) \cup \tilde{f}^{-1}\big(\mathcal{PG}(b + \tilde{u}'_b)\big)$ and $V_{\widetilde{\Theta}}$ intersect in 1 or 2 points. It follows that so do $\mathcal{PG}(a + \tilde{u}'_a) \cup \mathcal{PG}(b + \tilde{u}'_b)$ and $\tilde{f}(V_{\widetilde{\Theta}})$ and the result follows. □

Note that it is now fully established that all minimizing chains $(\tilde{\theta}_k)_{k\leqslant 0}$ calibrate a weak KAM solution thanks to Theorem 6.6.8 and Proposition 6.6.14. We are therefore allowed to use Corollary 6.6.4.

Follows an existence result of orbits displaying behaviors as in Corollary 6.6.15. The proof is inspired by a similar result for bi–infinite minimizing orbits in [30]. It shows that on each vertical above points that are not in the projected Mather set $\mathcal{M}_{\frac{p}{q}}$ there are at least two initial points of minimizing half orbits with rotation number p/q:

Proposition 6.6.17 *Assume that $\mathfrak{M}_{\rho_0} \neq \mathbb{R}$. Let (y_0^-, y_0^+) be a connected component of $\mathbb{R} \setminus \mathfrak{M}_{\rho_0}$, $(y_k^\pm)_{k\in\mathbb{Z}}$ be the associated minimizing orbits of type (p, q). Then for all $\tilde{\theta}_0 \in (y_0^-, y_0^+)$, there exists two minimizing chains $(\tilde{\theta}_k^\pm)_{k\leqslant 0}$ such that $\tilde{\theta}_0^\pm = \tilde{\theta}_0$ and such that $(\tilde{\theta}_k^+)_{k\leqslant 0}$ is α–asymptotic to $(y_k^+)_{k\leqslant 0}$ (resp. $(\tilde{\theta}_k^-)_{k\leqslant 0}$ is α–asymptotic to $(y_k^-)_{k\leqslant 0}$).*

Proof Let us prove the existence of $(\tilde{\theta}_k^-)_{k\leqslant 0}$, the other being obtained by reversing all inequalities.

Let $(\rho_n)_{n>0}$ be a decreasing sequence converging to p/q. For all $n > 0$, let $(\tilde{\theta}_k^n)_{k\leqslant 0}$ be a minimizing chain of rotation number ρ_n such that $\tilde{\theta}_k^n = \tilde{\theta}_0$ (for example a chain calibrating any weak KAM solution at a cohomology c_n verifying $\rho(c_n) = \rho_n$). By Corollary 6.6.4,

$$\forall n > 0, \quad \tilde{\theta}_q^n < \tilde{\theta}_0 + p.$$

By the same Corollary 6.6.4 or Proposition 6.6.1, for all $k \leqslant 0$, fixed, the sequence $(\tilde{\theta}_k^n)_{n>0}$ is bounded. Up to a diagonal extraction, we may therefore assume that for all $k \leqslant 0$, the sequence $(\tilde{\theta}_k^n)_{n>0}$ converges to some $\tilde{\theta}_k^-$. Clearly, the chain $(\tilde{\theta}_k^-)_{k\leqslant 0}$ is minimizing, as a limit of minimizing chains. Moreover, $\tilde{\theta}_0^- = \tilde{\theta}_0$ and $(\tilde{\theta}_k^-)_{k\leqslant 0}$ has rotation number p/q by passing to the limit in the inequalities provided by Proposition 6.6.1. Finally, $\tilde{\theta}_q^- \leqslant \tilde{\theta}_0 + p$. By Proposition 6.6.14, this cannot be an equality, hence $\tilde{\theta}_q^- < \tilde{\theta}_0 + p$ and by Corollary 6.6.15, $(\tilde{\theta}_k^-)_{k\leqslant 0}$ is α–asymptotic to $(y_k^-)_{k\leqslant 0}$. □

As a consequence, we recover a result of Mather [164] and Bangert [31]:

Theorem 6.6.18 *The following alternative holds:*

1. *the Mather set $\mathfrak{M}_{\frac{p}{q}} = \mathbb{R}$ in which case there is an invariant graph $C_{\frac{p}{q}} \subset \mathbb{A}$ on which the dynamics of f is q–periodic, there exists a unique weak KAM solution (up to constants) associated to the rotation number $\frac{p}{q}$ and the β function is derivable at $\frac{p}{q}$;*
2. *the Mather set $\mathfrak{M}_{\frac{p}{q}} \neq \mathbb{R}$ and the β function is not derivable at $\frac{p}{q}$.*

Proof

1. If $\mathfrak{M}_{\frac{p}{q}} = \mathbb{R}$, we infer from Proposition 3.5.1 that if $c \in \mathbb{R}$ verifies $\rho(c) = \frac{p}{q}$, and if u_c is a weak KAM solution at cohomology c, then u_c is derivable on $\mathbb{T}^1$ (hence C^1 and even $C^{1,1}$ by Birkhoff's Theorem6.1.3) and $c + u_c'$ does not depend on c, nor u_c, but only on the orbits of type (p, q), hence β is derivable at $\frac{p}{q}$ with $c = \beta'(p/q)$ and u_c is unique up to constants.
2. Assume now that $\mathfrak{M}_{\frac{p}{q}} \neq \mathbb{R}$. Let $[a, b] = \rho^{-1}(\{p/q\})$. Let u_a be the unique weak KAM solution at cohomology a that vanishes at 0 and u_b be the unique weak KAM solution at cohomology b that vanishes at 0. Let $\tilde{\theta}_0 \in \mathbb{R} \setminus \mathfrak{M}_{\frac{p}{q}}$ such that both $\tilde{u}_a$ and $\tilde{u}_b$ are derivable at $\tilde{\theta}_0$ (that exists as $\mathbb{R} \setminus \mathfrak{M}_{\frac{p}{q}}$ is a non–empty open set and $\tilde{u}_a$ and $\tilde{u}_b$ are derivable except on a countable set). Let finally $(\tilde{\theta}_k^{\pm})_{k\leqslant 0}$ be the sequences given by the previous Proposition 6.6.17. Then clearly, those two sequences are different, hence $\tilde{\theta}_{-1}^{+} \neq \tilde{\theta}_{-1}^{-}$ (more precisely, $\tilde{\theta}_{-1}^{-} < \tilde{\theta}_{-1}^{+}$). It follows from Proposition 6.6.14 and Theorem 6.3.8 that

$$a + \tilde{u}_a'(\tilde{\theta}_0) = \partial_2 \widetilde{S}(\tilde{\theta}_{-1}^{+}, \tilde{\theta}_0) \neq \partial_2 \widetilde{S}(\tilde{\theta}_{-1}^{-}, \tilde{\theta}_0) = b + \tilde{u}_b'(\tilde{\theta}_0).$$

   We deduce from Corollary 6.5.6 that $\tilde{\theta} \mapsto (\tilde{u}_b - \tilde{u}_a)(\tilde{\theta}) + (b - a)\tilde{\theta}$ is non–decreasing and non–constant. Finally, integrating inequality $b + \tilde{u}_b'(\tilde{\theta}) \geqslant a + \tilde{u}_a'(\tilde{\theta})$, that holds almost–everywhere, between 0 and 1 and remembering that it is not an equality almost–everywhere, yields $b > a$. As $[a, b] = \partial^- \beta(p/q)$ we have proven the result.

□

The previous proof actually implies the more precise result:

Proposition 6.6.19 *Assume $\mathcal{M}_{\frac{p}{q}} \neq \mathbb{T}^1$, let $[a, b] = \rho^{-1}(\{p/q\})$. Let u_a be the unique weak KAM solution at cohomology a that vanishes at 0 and u_b be the unique weak KAM solution at cohomology b that vanishes at 0. Then for all $\theta \in \mathbb{T}^1 \setminus \mathcal{M}_{\frac{p}{q}}$ where both u_a and u_b are derivable,*

$$a + u_a'(\theta) < b + u_b'(\theta).$$

As a consequence we deduce:

Proposition 6.6.20 *Assume that $\mathfrak{M}_{\rho_0} \neq \mathbb{R}$. Let (y_0^-, y_0^+) be a connected component of $\mathbb{R} \setminus \mathfrak{M}_{\rho_0}$, $(y_k^{\pm})_{k\in\mathbb{Z}}$ be the associated minimizing orbits of type (p, q). Then*

1. *all minimizing chains $(\tilde{\theta}_k^a)_{k\leqslant 0}$ calibrating $\tilde{u}_a$, with $\tilde{\theta}_0^a \in (y_0^-, y_0^+)$ are α–asymptotic to $(y_k^+)_{k\leqslant 0}$,*
2. *all minimizing chains $(\tilde{\theta}_k^b)_{k\leqslant 0}$ calibrating $\tilde{u}_b$, with $\tilde{\theta}_0^b \in (y_0^-, y_0^+)$ are α–asymptotic to $(y_k^-)_{k\leqslant 0}$.*

Proof Let us prove the first point. Let $(\tilde\theta_k^a)_{k\leqslant 0}$ be calibrating $\tilde u_a$. If $\tilde\theta_0^a$ is a point of derivability of $\tilde u_a$, let $(\tilde\theta_k^+)_{k\leqslant 0}$ be given by Proposition 6.6.17 that is α–asymptotic to $(y_k^+)_{k\leqslant 0}$ and such that $\tilde\theta_0^+ = \tilde\theta_0^a$. By Proposition 6.6.14 and Corollary 6.6.15 $(\tilde\theta_k^+)_{k\leqslant 0}$ calibrates $\tilde u_a$ and as the latter is derivable at $\tilde\theta_0^a$, such a minimizing chain is unique. Hence $(\tilde\theta_k^+)_{k\leqslant 0} = (\tilde\theta_k^a)_{k\leqslant 0}$ is indeed α–asymptotic to $(y_k^+)_{k\leqslant 0}$.

Let us now assume $\tilde\theta_0^a$ is not a point of derivability of $\tilde u_a$. Let $(\widetilde\Theta_0^n)_{n\geqslant 0}$ be an increasing sequence converging to $\tilde\theta_0^a$ and made of derivability points of $\tilde u_a$ (that is Lipschitz hence derivable almost everywhere). For all n, we denote by $(\widetilde\Theta_k^n)_{k\leqslant 0}$ the unique corresponding minimizing chain calibrating $\tilde u_a$. By the beginning of this proof, for n fixed, $(\widetilde\Theta_k^n)_{k\leqslant 0}$ is α–asymptotic to $(y_k^+)_{k\leqslant 0}$ and verifies $\widetilde\Theta_q^n > \widetilde\Theta_0^n + p$ by Corollary 6.6.15. Moreover, setting $r_k^n = a + \tilde u_a'(\widetilde\Theta_k^n)$, $(\widetilde\Theta_k^n, r_k^n)_{k\leqslant 0}$ is a piece of orbit of $\tilde f$ by Theorem 6.3.8. It follows that as $n \to +\infty$, $(\widetilde\Theta_k^n, r_k^n)_{n\leqslant 0}$ converges to $(\tilde\theta_k^{a-}, r_k^-)$ such that $(\tilde\theta_k^{a-}, r_k^-)_{k\leqslant 0}$ is a piece of orbit of $\tilde f$, $\tilde\theta_0^{a-} = \tilde\theta_0^a$ and $r_0^- = a + u_{a-}'(\tilde\theta_0^a)$. Moreover, $(\tilde\theta_k^{a-})_{k\leqslant 0}$ calibrates $\tilde u_a$. By passing to the limit in the corresponding inequalities for $\widetilde\Theta^n$ we gather $\tilde\theta_q^{a-} \geqslant \tilde\theta_0^{a-} + p$ and equality is excluded by Proposition 6.6.14, hence $\tilde\theta_q^{a-} > \tilde\theta_0^{a-} + p$ and $(\tilde\theta_k^{a-})_{k\leqslant 0}$ is α-asymptotic to $(y_k^+)_{k\leqslant 0}$.

Coming back to the chain $(\tilde\theta_k^a)_{k\leqslant 0}$, setting $r_0^a = \partial_2\widetilde S(\tilde\theta_{-1}^a, \tilde\theta_0^a)$, as $r_0^a - a \in \partial^+\tilde u_a(\tilde\theta_0^a)$ and by semiconcavity of $\tilde u_a$, it follows that $r_0^a \leqslant r_0^-$. Then, by the twist condition, $\tilde\theta_{-1}^a \geqslant \tilde\theta_{-1}^{a-}$. Last, applying again that calibrating chains do not cross away from the origin (Lemma 6.3.12) gives $\tilde\theta_k^a \geqslant \tilde\theta_k^{a-}$ for all $k \geqslant 0$. We then apply to $k = q$ in order to conclude that

$$\tilde\theta_q^a \geqslant \tilde\theta_q^{a-} > \tilde\theta_0^{a-} + p = \tilde\theta_0^a + p.$$

The result now follows from Proposition 6.6.14. □

The proof of Theorem 6.6.13 is now fully completed.

To clarify the picture the following result makes Proposition 6.6.19 more precise. As a matter of fact, it implies that the full pseudographs $\mathcal{PG}(a+u_a')$ and $\mathcal{PG}(b+u_b')$ only intersect on the Mather set.

Proposition 6.6.21 *Assume that $\mathfrak{M}_{\rho_0} \neq \mathbb{R}$. Let (y_0^-, y_0^+) be a connected component of $\mathbb{R}\setminus\mathfrak{M}_{\rho_0}$. Then restricted to (y_0^-, y_0^+), $\mathcal{PG}(a+\tilde u_a')$ is strictly under $\mathcal{PG}(b+\tilde u_b')$, in the sense that if $\tilde\theta_0 \in (y_0^-, y_0^+)$ and r_a is such that $(\tilde\theta_0, r_a) \in \mathcal{PG}(a+\tilde u_a')$ and r_b is such that $(\tilde\theta_0, r_b) \in \mathcal{PG}(b+\tilde u_b')$, then $r_a < r_b$.*

Proof Let $(y_k^\pm)_{k\in\mathbb{Z}}$ be the minimizing orbits of rotation number ρ_0 that are of type (p, q). Then as the sequence $(y_k^+ - y_k^-)_{k\in\mathbb{Z}}$ is positive valued and $|q|$–periodic, there exists $\varepsilon > 0$ such that $y_k^- - y_k^+ > \varepsilon$ for all integer $k \in \mathbb{Z}$.

Let $\tilde\theta_0 \in (y_0^-, y_0^+)$ and let us assume by contradiction that there exist two minimizing chains $(\tilde\theta_k^a)_{k\leqslant 0}$ and $(\tilde\theta_k^b)_{k\leqslant 0}$ calibrating respectively $\tilde u_a$ and $\tilde u_b$, such that $\tilde{\theta}_0^a = \tilde{\theta}_0^b = \tilde{\theta}_0 \in (y_0^-, y_0^+)$ and verifying $r_0^a \geqslant r_0^b$, where $r_0^a = \partial_2\widetilde{S}(\tilde{\theta}_{-1}^a, \tilde{\theta}_0^a)$ and $r_0^b = \partial_2\widetilde{S}(\tilde{\theta}_{-1}^b, \tilde{\theta}_0^b)$. First note that $r_0^a \neq r_0^b$ as otherwise both minimizing chains would be equal and $(\tilde{\theta}_k^a)_{k\leqslant 0}$ is α–asymptotic to $(y_k^+)_{k\leqslant 0}$ while $(\tilde{\theta}_k^b)_{k\leqslant 0}$ is α–asymptotic to $(y_k^-)_{k\leqslant 0}$. As now $r_0^a > r_0^b$, it follows from the twist condition that $\tilde{\theta}_{-1}^a < \tilde{\theta}_{-1}^b$. But for $n < 0$ large enough in absolute value, $|\tilde{\theta}_n^a - y_n^+| < \frac{\varepsilon}{2}$ and $|\tilde{\theta}_n^b - y_n^-| < \frac{\varepsilon}{2}$ thus implying that $\tilde{\theta}_n^a > \tilde{\theta}_n^b$. Hence the two minimizing chains cross at 0 and somewhere between -1 and n, that is absurd.

The previous result applies in particular to the minimizing and calibrating chains verifying $r_0^a = a + \tilde{u}'_{a-}(\tilde{\theta}_0)$ and $r_0^b = b + \tilde{u}'_{b+}(\tilde{\theta}_0)$ therefore proving the Proposition. □

Now that we have a pretty good idea of how are organized the extreme pseudographs $\mathcal{PG}(a + u'_a)$ and $\mathcal{PG}(b + u'_b)$ and their respective calibrating chains, the next results help understanding the looks of more general weak KAM solutions at any cohomology class $c \in (a, b)$.

Proposition 6.6.22 *Assume that* $\mathfrak{M}_{\rho_0} \neq \mathbb{R}$. *Let* (y_0^-, y_0^+) *be a connected component of* $\mathbb{R} \setminus \mathfrak{M}_{\rho_0}$. *Let* $c \in (a, b)$, $v_c : \mathbb{T}^1 \to \mathbb{R}$ *a weak KAM solution at cohomology* c *and* $\tilde{\theta}_0 \in (y_0^-, y_0^+)$ *a point of derivability of* $\tilde{v}_c$ *the lift of* v_c. *Then one of the following holds:*

1. $\tilde{u}_a$ *is derivable at* $\tilde{\theta}_0$ *and* $a + \tilde{u}'_a(\tilde{\theta}_0) = c + \tilde{v}'_c(\tilde{\theta}_0)$;
2. $\tilde{u}_b$ *is derivable at* $\tilde{\theta}_0$ *and* $b + \tilde{u}'_b(\tilde{\theta}_0) = c + \tilde{v}'_c(\tilde{\theta}_0)$.

Proof Let $(\tilde{\theta}_k^c)_{k\leqslant 0}$ be the unique minimizing chain calibrating $\tilde{v}_c$ such that $\tilde{\theta}_0^c = \tilde{\theta}_0$. By Theorem 6.6.13, either $\tilde{\theta}_0^c < \tilde{\theta}_q^c - p$ and $(\tilde{\theta}_k^c)_{k\leqslant 0}$ calibrates $\tilde{u}_a$, either $\tilde{\theta}_0^c > \tilde{\theta}_q^c - p$ and $(\tilde{\theta}_k^c)_{k\leqslant 0}$ calibrates $\tilde{u}_b$.

Let us consider the first case and prove that $\tilde{u}_a$ is derivable at $\tilde{\theta}_0$. Recall that as $\tilde{v}_c$ is a semiconcave function that is derivable at $\tilde{\theta}_0$, then $\tilde{v}'_c$ is continuous at $\tilde{\theta}_0$. Let $\mathcal{D} \subset \mathbb{R}$ be a set, the complement of which is countable, such that all functions $\tilde{u}_a$, $\tilde{u}_b$ and $\tilde{v}_c$ are derivable on $\mathcal{D}$. By continuity of $\tilde{v}'_c$ at $\tilde{\theta}_0$, there exists $\varepsilon > 0$ such that for all $\widetilde{\Theta}_0 \in \mathcal{D} \cap (\tilde{\theta}_0 - \varepsilon, \tilde{\theta}_0 + \varepsilon) \subset (y_0^-, y_0^+)$,

$$\pi_1 \circ \tilde{f}^q\big(\widetilde{\Theta}_0, c + \tilde{v}'_c(\widetilde{\Theta}_0)\big) - p > \widetilde{\Theta}_0,$$

where we use that $\tilde{\theta}_q^c = \pi_1 \circ \tilde{f}^q\big(\tilde{\theta}_0, c + \tilde{v}'_c(\tilde{\theta}_0)\big)$. It follows that, setting $\widetilde{\Theta}_k = \pi_1 \circ \tilde{f}^k\big(\widetilde{\Theta}_0, c + \tilde{v}'_c(\widetilde{\Theta}_0)\big)$ for all $k \leqslant 0$, the chain $(\widetilde{\Theta}_k)_{k\leqslant 0}$ is the unique calibrating chain for $\tilde{v}_c$ starting at $\widetilde{\Theta}_0$ and that it also calibrates $\tilde{u}_a$, again using Theorem 6.6.13. As, by definition of $\mathcal{D}$, such a calibrating chain is also unique for $\tilde{u}_a$ we uncover that

$$a + \tilde{u}'_a(\widetilde{\Theta}_0) = c + \tilde{v}'_c(\widetilde{\Theta}_0) = \partial_2\widetilde{S}(\widetilde{\Theta}_{-1}, \widetilde{\Theta}_0).$$

Applying the preceding equality to an increasing sequence $(\widetilde{\Theta}_0^n)_{n\geqslant 0}$ of points in $\mathcal{D} \cap (\tilde{\theta}_0 - \varepsilon, \tilde{\theta}_0 + \varepsilon)$ converging to $\tilde{\theta}_0$ gives $a + \tilde{u}'_{a-}(\tilde{\theta}_0) = c + \tilde{u}'_c(\tilde{\theta}_0)$. Similarly, taking a decreasing sequence $(\widetilde{\Theta}_0^n)_{n\geqslant 0}$ of points in $\mathcal{D} \cap (\tilde{\theta}_0 - \varepsilon, \tilde{\theta}_0 + \varepsilon)$ converging to $\tilde{\theta}_0$ gives $a + \tilde{u}'_{a+}(\tilde{\theta}_0) = c + \tilde{u}'_c(\tilde{\theta}_0)$. Those two equalities provide the desired result. □

We deduce from Propositions 6.6.21 and 6.6.22 and from semiconcavity, that weak KAM solutions' pseudographs can jump downward only once from $\mathcal{PG}(b + \tilde{u}'_b)$ to $\mathcal{PG}(a + \tilde{u}'_a)$ on each connected component of $\mathbb{R} \setminus \mathfrak{M}_{\rho_0}$.

Theorem 6.6.23 *Assume that* $\mathfrak{M}_{\rho_0} \neq \mathbb{R}$. *Let* (y_0^-, y_0^+) *be a connected component of* $\mathbb{R} \setminus \mathfrak{M}_{\rho_0}$. *Let* $c \in (a, b)$, $v_c : \mathbb{T}^1 \to \mathbb{R}$ *a weak KAM solution at cohomology c. Then there exists* $\widetilde{\Theta} \in [y_0^-, y_0^+]$ *such that*

1. $c + \tilde{v}'_c(\tilde{\theta}) = b + \tilde{u}'_b(\tilde{\theta})$ *for almost every* $\tilde{\theta} \in (y_0^-, \widetilde{\Theta})$,
2. $c + \tilde{v}'_c(\tilde{\theta}) = a + \tilde{u}'_a(\tilde{\theta})$ *for almost every* $\tilde{\theta} \in (\widetilde{\Theta}, y_0^+)$,
3. $c + \tilde{v}'_{c-}(\widetilde{\Theta}) = b + \tilde{u}'_{b-}(\widetilde{\Theta})$,
4. $c + \tilde{v}'_{c+}(\widetilde{\Theta}) = a + \tilde{u}'_{a+}(\widetilde{\Theta})$.

In particular,

$$\forall \tilde{\theta} \in (y_0^-, \widetilde{\Theta}), \quad \tilde{v}_c(\tilde{\theta}) = \tilde{v}_c(y_0^-) + \int_{y_0^-}^{\tilde{\theta}} \tilde{u}'_b(s)\, \mathrm{d}s + (b - c)(\tilde{\theta} - y_0^-);$$

$$\forall \tilde{\theta} \in (\widetilde{\Theta}, y_0^+), \quad \tilde{v}_c(\tilde{\theta}) = \tilde{v}_c(y_0^+) + \int_{y_0^+}^{\tilde{\theta}} \tilde{u}'_a(s)\, \mathrm{d}s + (a - c)(\tilde{\theta} - y_0^+).$$

6.7 A Glimpse into the World of Weakly Integrable Twist Maps

We wish to give an account on some results originally published in [21, 22] by Arnaud–Zavidovique. We will only state them as the proofs go far beyond the scope of this memoir. The understanding of weakly integrable twist maps (Definition 6.2.7) is a frustrating task. Indeed, as was already pointed out, there is no known example of C^0–integrable twist map with a non C^1 invariant circle. The notion appears in various historic works (even if not explicitly defined). In the study of the Hopf conjecture about Riemannian tori without conjugate points, before its definitive answer by Burago and Ivanov in [61], it was proved by Heber [128] that such geodesic flows are C^0–integrable. This result was then generalized to exact symplectic twist maps [80] and at last to more general flows of Tonelli Hamiltonians ([4] and also [11]) and twist maps in higher dimension [3]. Finally, on more general surfaces, let us mention the work [158].

The philosophy of our results is to show that weak forms of integrability have strong dynamical implications and that further properties of the underlying foliations can be obtained. The first result completely characterizes C^0–integrable twist maps in terms of the function $u : \mathbb{T}^1 \times \mathbb{R} \to \mathbb{R}$ provided by Theorem 6.5.3.

Theorem 6.7.1 *There is equivalence between*

1. *the map f is C^0–integrable,*
2. *the function u is C^1.*

Moreover in either case the function u is unique and

- *for each $c \in \mathbb{R}$, the graph $\mathcal{G}(c + u'_c)$ is a leaf of the invariant foliation $\mathcal{F}$,*
- *the map $h_c : \theta \mapsto \theta + \frac{\partial u}{\partial c}(\theta, c)$ is a semi–conjugation between the projected dynamics $g_c : \theta \mapsto \pi_1 \circ f\big(\theta, c + \frac{\partial u}{\partial \theta}(\theta, c)\big)$ and the rotation $R_{\rho(c)} : \theta \mapsto \theta + \rho(c)$ meaning that $h_c \circ g_c = R_{\rho(c)} \circ h_c$.*

The striking fact in the previous Theorem is the regularity with respect to c. At irrational rotation numbers, Poincaré–Denjoy theory gives that a semi–conjugation to the corresponding rotation is unique and regularity at such cohomology classes is not surprising. It is not the case at cohomology classes with a rational rotation number and the proof actually gives that the function h_c is C^{k-1} if f is C^k. Moreover, previous works of Arnaud [6] yield that at such a cohomology class c, the invariant circle is also C^k (that also follows from the Implicit Function Theorem) and the restricted dynamics is completely periodic and conjugated to a rotation (cf. Theorem 6.6.18).

The main (hypothetical) feature of a C^0–integrable twist map that would not be integrable is the presence of invariant circles with irrational rotation number and a restricted dynamics that is one of a Denjoy counterexample. This means that if c is the corresponding cohomology class, the map h_c is not a homeomorphism. This is excluded in the case of Lipschitz–integrable twist maps (see Definition 6.2.6) by the next Theorem.

Theorem 6.7.2 *Assume that the Exact Conservative Twist Map is Lipschitz integrable. Then there exists an exact area preserving homeomorphism Φ of $\mathbb{T}^1 \times \mathbb{R}$, which is C^1 in the variable θ, such that*

$$\forall (x, c) \in \mathbb{T}^1 \times \mathbb{R}, \quad \Phi \circ f \circ \Phi^{-1}(x, c) = (x + \rho(c), c).$$

Moreover, in this case, $\rho : \mathbb{R} \to \mathbb{R}$ is a bi–Lipschitz homeomorphism, all the leaves of the invariant foliation are C^1 and the restricted dynamics on each leaf is C^1–conjugated to a rotation.

The function Φ is implicitly defined by the relation

$$\Phi\Big(\theta, c + \frac{\partial u}{\partial \theta}(\theta, c)\Big) = \Big(\theta + \frac{\partial u}{\partial c}(\theta, c), c\Big).$$

In the previous Theorem, the area preserving homeomorphism Φ maps the foliation $\mathcal{F}$ invariant by f to the standard foliation $\mathcal{F}^0$ consisting of the obvious circles $\mathcal{F}^0_c = \{(\theta, c),\ \theta \in \mathbb{T}^1\}$. A natural problem is therefore to find which foliations by graphs are homeomorphic to the standard foliation by an exact area preserving homeomorphism. On this matter, we provide the following characterization.

Theorem 6.7.3 *Let $\mathcal{F}$ be a foliation of $\mathbb{T}^1 \times \mathbb{R}$ by graphs of functions $\theta \mapsto \eta_c(\theta)$ such that for $c \in \mathbb{R}$, $\int_{\mathbb{T}^1} \eta_c(\theta)\, \mathrm{d}\theta = c$. Then $\mathcal{F}$ is homeomorphic to the standard foliation $\mathcal{F}^0$ by an exact area preserving homeomorphism if and only if there exists a C^1 function $u : \mathbb{T}^1 \times \mathbb{R} \to \mathbb{R}$ such that*

- $u(0, c) = 0$ *for all* $c \in \mathbb{R}$,
- $\eta_c(\theta) = c + \frac{\partial u}{\partial \theta}(\theta, c)$ *for all* $(\theta, c) \in \mathbb{T}^1 \times \mathbb{R}$,
- *for all* $c \in \mathbb{R}$, *the map* $\theta \mapsto \theta + \frac{\partial u}{\partial c}(\theta, c)$ *is a homeomorphism of* $\mathbb{T}^1$.

Again, an area preserving homeomorphism Φ sending $\mathcal{F}$ to $\mathcal{F}^0$ is implicitly defined by the relation

$$\Phi\Big(\theta, c + \frac{\partial u}{\partial \theta}(\theta, c)\Big) = \Big(\theta + \frac{\partial u}{\partial c}(\theta, c), c\Big).$$

As a conclusion, the previous Theorem allows to explain Theorem 6.2.8. In the case of the foliation given by the functions $\eta_c(\theta) = c + \varepsilon(c)\cos(2\pi\theta)$, for a function $\varepsilon : \mathbb{R} \to \mathbb{R}$ which is Lipschitz, non C^1, with a small enough Lipschitz constant, the foliation is Lipschitz in the sense of Definition 6.2.6. Moreover, were this foliation straightened by an exact area preserving homeomorphism the associated function u would be given by

$$\forall (\theta, c) \in \mathbb{T}^1 \times \mathbb{R}, \quad u(\theta, c) = \frac{\varepsilon(c)}{2\pi} \sin(2\pi\theta).$$

This last function is clearly not C^1, thus violating the conclusion of Theorem 6.7.3.

We conclude that the foliation given by η cannot be invariant by an ECTM.

Epilogue: Weak KAM Theory, Origins and Present Developments

To end this text, we will develop the reasons that led Albert Fathi to name "Weak KAM" his theory and present some of its more recent developments. The choices made by the author are by no means exhaustive, as the subject is very active and researchers quite prolific.

Etymology of Weak KAM

In order to understand the origin of the weak KAM terminology, one has to understand the KAM theorem. KAM is the acronym for Kolmogorov Arnol'd Moser. Briefly, Kolmogorov [148] was the first to state a KAM like theorem in an analytic setting. Then Arnol'd [24] was the first to prove the theorem stated by Kolmogorov (and others in the same spirit during his career). He also extended it in celestial mechanics [25] providing an incomplete proof later made rigorous in [119] by Herman and Fejoz. Finally Moser [177] extended those theorems to C^∞ and finite differentiability settings.

Following closely [120] let us state the invariant torus theorem of Kolmogorov. Let $\mathcal{H}$ be the set of Hamiltonians on $\mathbb{T}^N \times \mathbb{R}^N$ that are real analytic on a neighborhood of the 0–section $T_0 = \mathbb{T}^N \times \{0\}$. For $\alpha \in \mathbb{R}^N$, let $\mathcal{K}^\alpha$ be the set of Hamiltonians $K \in \mathcal{H}$ that are of the form

$$K(\theta, r) = c + \langle \alpha, r\rangle + O(r^2),$$

where $\langle \cdot, \cdot\rangle$ is the canonical scalar product on $\mathbb{R}^N$, $c \in \mathbb{R}$ is a constant and $O(r^2)$ is a function of (r, θ) such that $O(r^2)/\|r\|^2$ is bounded as $r \to 0$. In particular, this implies that T_0 is invariant by the Hamiltonian flow ϕ_K of K and that the Hamiltonian vector–field is constantly equal to $(\alpha, 0)$ when restricted to T_0. In

M. Zavidovique, *Discrete Weak KAM Theory*, Lecture Notes in Mathematics 2377,
https://doi.org/10.1007/978-3-031-96809-9

particular, $\phi_K^t(\theta, 0) = (\theta + t\alpha, 0)$ for all $t \in \mathbb{R}$ and $\theta \in \mathbb{T}^N$ meaning that ϕ_K restricted to T_0 is a translation flow of vector α.

Let $\mathcal{G}$ be the space of symplectic transformations G of $\mathbb{T}^N \times \mathbb{R}^N$ that are real analytic in a neighborhood of T_0 and of the form

$$G(\theta, r) = \Big(\varphi(\theta), (D_\theta\varphi)^{-1}\big(r + \rho(\theta)\big)\Big),$$

where $\varphi : \mathbb{T}^N \to \mathbb{T}^N$ is analytic with $\varphi(0) = 0$, and $\rho(\theta)d\theta$ is a closed 1– form on $\mathbb{T}^N$.

Now we focus on $\alpha \in \mathbb{R}^N$ that is Diophantine meaning that there is a small $0 < \gamma \ll 1$ and $\tau > N - 1$ such that

$$|\langle k, \alpha\rangle| \geqslant \gamma \|k\|_1^{-\tau},$$

for all $k \in \mathbb{Z}^N \setminus \{0\}$, where $\|k\|_1 = |k_1| + \cdots + |k_N|$. We start from a Hamiltonian $K^0 \in \mathcal{K}^\alpha$ that has Taylor expansion

$$K^0(\theta, r) = c^0 + \langle \alpha, r\rangle + Q_\theta^0(r, r) + O(r^3)$$

where Q_θ^0 is a symmetric bilinear form for all $\theta \in \mathbb{T}^N$ with a non–degeneracy condition $\det\Big(\int_{\mathbb{T}^N} Q_\theta^0 \, \mathrm{d}\theta\Big) \neq 0$. In this setting the KAM theorem then states:

Theorem *For every $H \in \mathcal{H}$ close to K^0, there exists $(K, G) \in \mathcal{K}^\alpha \times \mathcal{G}$ close to (K^0, Id) such that $H = K \circ G$ in a neighborhood of $G^{-1}(T_0)$.*

In the previous Theorem, the notions of proximity are in the analytic sense. The Theorem means that if K^0 has the 0–section as invariant manifold with a suitable restricted dynamics (a Diophantine translation) then a small perturbation H of K^0 has an invariant torus $G^{-1}(T_0)$, called a KAM torus, close to the initial invariant torus T^0. Moreover the restricted dynamics of H on $G^{-1}(T_0)$ is conjugated to the same Diophantine translation. The previous theorem will also clearly hold true by composing by an analytic symplectomorphism.

If now H is sufficiently close to K^0, its invariant torus $G^{-1}(T_0)$ will still be a graph of a closed 1–form meaning that there is $P \in \mathbb{R}^N$ (a cohomology class) and $u : \mathbb{T}^N \to \mathbb{R}$ smooth such that $G^{-1}(T_0) = \{(\theta, P + D_\theta u), \;\; \theta \in \mathbb{T}^N\}$. We now invoke basic results from symplectic geometry stating that the graph of a closed 1–form is a Lagrangian submanifold and that a Lagrangian submanifold $\mathcal{L}$ is invariant by a Hamiltonian flow ϕ_H if and only if H is constant on $\mathcal{L}$ to infer that

$$\exists C \in \mathbb{R}, \quad \forall \theta \in \mathbb{T}^N, \quad H(\theta, P + D_\theta u) = C.$$

The main output we will keep from the deep KAM theorem is that a KAM torus yields a smooth strong solution of a stationary Hamilton–Jacobi equation. On the contrary, as we already extensively developed, weak KAM solutions are viscosity solutions of stationary Hamilton–Jacobi equations and viscosity solutions are a form of weak solutions. This is the etymology:

strong KAM theory provides strong solutions of the Hamilton–Jacobi equation whereas weak KAM theory provides weak solutions of the Hamilton–Jacobi equation.

In the particular case of KAM tori, classical objects from Aubry–Mather theory detailed in this book are studied in [190].

Arnol'd Diffusion

Following the previous paragraph, assume we start from an analytic Hamiltonian that only depends on the momentum variable: $K^0 : \mathbb{T}^N \times \mathbb{R}^N$ such that $K^0(\theta, r) = k^0(r)$ where $k^0 : \mathbb{R}^N \to \mathbb{R}$ is an analytic non constant function. Writing the Hamiltonian vector–field $X_{K^0}(\theta, r) = (\nabla k_0(r), 0)$ shows that each torus $T_r = \mathbb{T}^N \times \{r\}$ is invariant by the Hamiltonian flow and the restricted dynamics on T_r is a translation of vector $\nabla k_0(r)$. The KAM theorem asserts that if H is close to K^0 then H will possess many invariant tori that it inherits from K^0. A natural idea is then that those invariant tori of H will prevent other orbits from wandering much in the phase space. Arnol'd [26] exhibits examples showing the contrary.

For $\varepsilon > 0$ and $\mu > 0$ let

$$H_{\varepsilon,\mu}(t, \theta, q, I, p) = \frac{1}{2}|p|^2 + \frac{1}{2}|I|^2 + \varepsilon(\cos(2\pi q) - 1) \\ + \mu(\cos(2\pi q) - 1)\big(\sin(2\pi\theta) + \cos(2\pi t)\big).$$

We see $H_{\varepsilon,\mu}$ as a time periodic Hamiltonian defined on $\mathbb{R} \times \mathbb{T}^2 \times \mathbb{R}^2$. For $\varepsilon = \mu = 0$, $H_{0,0}$ is an integrable Hamiltonian as described previously. If now $\varepsilon > 0$ is fixed, the Hamiltonian flow of $H_{\varepsilon,0}$ is still easy to understand (product of a rotator and a pendulum) and its trajectories $\big(\theta(t), q(t), I(t), p(t)\big)$ verify $I(t) = cste$. One could then expect $I(t)$ not to vary too much for small values of $\mu > 0$ but Arnol'd proves the following:

Theorem *For any* $0 < I_- < I_+$ *and* $\mu > 0$ *small enough, there is* $T > 0$ *and a solution* $t \mapsto \big(\theta(t), q(t), I(t), p(t)\big)$ *of the Hamiltonian flow of* $H_{\varepsilon,\mu}$ *such that* $I(0) < I_-$ *and* $I(T) > I_+$.

For several decades following this example, it has been a huge challenge to extend it to more general perturbations of integrable (or not) systems and even maybe establish the genericity of such phenomenons. As we already pointed out, Mather developed his theory to construct diffusion orbits [167, 169, 171]. Many tried to implement Mather's ideas and Arnol'd's mechanism to reach the same goal, for instance Bernard combined Mather's ideas with weak KAM theory in [43].

Probably the most striking breakthrough was achieved by Bernard et al. [55], following preparatory work of Bernard [44] mixing weak KAM, Aubry-Mather and shadowing of hyperbolic objects. They prove the following, starting from a uniformly convex Hamiltonian $H_0 : B_N \to \mathbb{R}$ of class C^r, $r \geqslant 4$, where B_N is the unit ball of $\mathbb{R}^N$. Denoting $\mathcal{S}^r$ the unit sphere in $C^r(\mathbb{T} \times \mathbb{T}^N \times B_N)$:

Theorem *There exist two continuous functions ℓ and ε_0 on $\mathcal{S}^r$ that are positive on an open and dense set $\mathcal{U} \subset \mathcal{S}^r$ and an open and dense subset*

$$\mathcal{V}_1 \subset \{H_0 + \varepsilon H_1,\ H_1 \in \mathcal{U},\ 0 < \varepsilon < \varepsilon_0(H_1)\}$$

such that it holds:

If $H \in \mathcal{V}_1$, there exists an orbit (θ, p) of the Hamiltonian flow of H_ε and $T \in \mathbb{N}$ such that $\|p(T) - p(0)\| \geqslant \ell(H_1)$.

In the previous theorem the crucial point is that even if the diffusion is not arbitrarily large, it is of a quantity $\ell(H_1)$ that does not depend on ε. This result was then improved in two and a half degrees of freedom, achieving Mather's goal, in [144].

The Mañé Conjecture and the Size of the Aubry Set

Another important genericity problem is the Mañé conjecture. In [154], Mañé proves that for a generic Tonelli Hamiltonian H on a compact manifold (for a notion now called Mañé genericity), H only has one Mather measure. He asks questions to improve this result, one being the famous Mañé conjecture that we reproduce here:

Problem III Is it true that for a generic Tonelli Lagrangian L, there exists an open and dense set $\mathcal{H} \subset H^1(M, \mathbb{R})$ such that $\mathcal{M}^\omega(L)$ consists of a unique minimizing measure supported in a periodic orbit [if $\omega \in \mathcal{H}$]?

In this Problem, $\mathcal{M}^\omega(L)$ is the set of Mather measures at cohomology ω. This conjecture has been often restated in an almost equivalent one: is it true that for a Mañé generic Lagrangian L there is a unique Mather measure supported on a hyperbolic periodic orbit.

In the context of Arnol'd diffusion, supports of Mather measures, as well as Aubry sets and Mather sets, being invariant, are more of an obstruction to the existence of diffusing orbits. It is therefore not a surprise that Mather also was interested in the generic size of those sets (see [168, 170]). In particular he conjectured that for a generic Tonelli Lagrangian, $\mathcal{M}^\omega(L)$ contains a finite number of ergodic measures for all $\omega \in \mathcal{H} \subset H^1(M, \mathbb{R})$. He also posed questions about the total disconnectedness of the quotiented Aubry set, giving examples where it is not (see references aforementioned).

Concerning Mather's conjecture, it was solved in [46, 50] where for example it is proved that

Theorem *Let $L : \mathbb{T} \times M \to \mathbb{R}$ be a smooth Tonelli Hamiltonian then there exists a residual subset O of $C^\infty(\mathbb{T} \times M, \mathbb{R})$ such that for all $u \in O$ and $c \in H^1(M, \mathbb{R})$ there are at most $1 + dim\ H^1(M, \mathbb{R})$ ergodic minimizing measures of $L - u - c$.*

Concerning the total disconnectedness problem, let us mention [189] and [118] where Sard like theorems are used. For example in the later reference, amongst much more precise results, the authors prove that

Theorem *For an autonomous smooth Tonelli Lagrangian $L : TM \to \mathbb{R}$, if M is of dimension $d \leqslant 3$, the one dimensional Hausdorff measure of the quotiented Aubry set is 0.*

Concerning Mañé's conjecture, it is still open to this day but substantial progress has been made. Following first results by Figalli and Rifford [121, 122] where the authors assume for example the existence of critical subsolutions with certain regularity, an important step was made in [84]:

Theorem *Let $H : T^*M \to \mathbb{R}$ be a Tonelli Hamiltonian of class C^2, and assume that the dimension of M is 2. Then there is an open dense set of potentials $\mathcal{V} \subset C^2(M, \mathbb{R})$ such that, for every $V \in \mathcal{V}$, the Aubry set associated to the Hamiltonian $H + V$ is hyperbolic in its energy level.*

Hyperbolocity of the Aubry set is a great step as it is an open property. Following this, a recent preprint of the last author [83] claims the proof of Mañé's conjecture in dimension 2, in the C^2 topology. This work is still under review when those lines are being written. Let us mention that an important result of similar flavor had been obtained by the same author [82]. This day, we sadly learned that Gonzalo Contreras passed away prematurely. We seize this opportunity to pay tribute to a great mathematician who leaves an important legacy in dynamical systems and in Mathematics more generally.

The Shape of the Aubry Set

As is well known, not only size matters. In a sensational series of papers [5–7, 9, 13, 14, 17, 18], Marie–Claude Arnaud established surprising results linking the dynamics on the Aubry sets (hyperbolicity, Lyapunov exponents, angles between Oseledet's splitting) and its shape (regularity). Her results concern both Tonelli Lagrangian systems, and symplectic twist maps of higher dimensional annuli.

The main tool of those results is the notion of Green bundles G_+ and G_- that are defined along the Aubry set. They are invariant semicontinuous bundles that trap the Aubry set. Therefore, according to how the Green bundles intersect, Arnaud deduces either regularity properties of the Aubry set, or of supports of Mather measures, or she finds hyperbolicity properties for the restricted dynamics. Far from being fair to the huge variety of results Arnaud proves, let us mention two. In [6] she finds properties of invariant essential curves of ECTM of the annulus. The Birkhoff Theorem asserts that they are Lipschitz graphs.

Theorem *Let $f : \mathbb{A} \to \mathbb{A}$ be an exact symplectic C^1 positive twist map and let $\gamma : \mathbb{T} \to \mathbb{R}$ be a Lipschitz map whose graph is invariant by f. Then there exists a dense G_δ subset U of $\mathbb{T}$ whose Lebesgue measure is 1 and such that every t of U is a point of differentiability of γ and a point of continuity of γ'.*

This allows to exhibit graphs of Lipschitz functions that cannot be invariant by an ECTM.

In [13] this Theorem is suitably extended to Tonelli Lagrangian systems. For the dynamical aspects Arnaud proves:

Theorem *Let M be an n dimensional manifold and $H : T^*M \to \mathbb{R}$ be a Tonelli Hamiltonian and let μ be an ergodic minimizing probability measure. Then the two following assertions are equivalent:*

- *at μ–almost every point,* $\dim\big(G_-(x) \cap G_+(x)\big) = p$;
- *μ has exactly $2p$ zero Lyapunov exponents, $n-p$ positive ones and $n-p$ negative ones.*

Singularities of Weak KAM Solutions

As already hinted in the previous sections, regularity of subsolutions and of weak KAM solutions can provide interesting consequences on the Aubry, Mather sets and on the Hamiltonian dynamics. We already mentioned that for a Tonelli Hamiltonian, the weak KAM solutions are Lipschitz hence differentiable almost everywhere. They even are locally semiconcave and it can be proved that if the dimension of M is n, the set of singularities (non–differentiability points) of a semiconcave function is $(n-1)$–rectifiable. This means that it is included in a countable union of Lipschitz submanifolds (see [67]). Better understanding the set of singularities of weak KAM solutions, or more generally of solutions of the time evolution Hamilton–Jacobi equation, is an important and long standing challenge. It was for example conjectured that there is a sort of intrinsic dynamics taking place inside the set of singularities. Such problems were for instance studied in [68, 69], see also [66]. In [65] the authors had the excellent idea to use (yet again!) the positive Lax–Oleinik semigroup to trace back singularities of solutions to the negative Lax–Oleinik semigroup. This idea was then isolated and systematized in [70, 72]. In the first paper, the authors use this to construct homotopy equivalences between sets involving the Aubry set of a weak KAM solution u and the set of its singularities. In the second paper, they extend those results and apply them to Riemannian geometry problems and singularities of distance functions. Interesting results in this line of research had been obtained in [2] and are recovered and widely extended. We will give one striking example. If (M, g) is a complete Riemannian manifold let

$\mathcal{NU}(M, g)$ be the set of pairs $(x, y) \in M \times M$ that are joined by at least two distinct minimizing geodesics. We denote $\Delta_M = \{(x, x), \quad x \in M\}$.

Theorem *The set $\mathcal{NU}(M, g) \subset M \times M \setminus \Delta_M$ is locally contractible. In particular $\mathcal{NU}(M, g)$ is locally path connected.*

Contact Type Hamilton–Jacobi Equations

As already mentioned in the introduction, contact type Hamiltonians are defined on the 1–jet bundle $T^*M \times \mathbb{R}$ of a closed manifold M. The name comes from the fact that $T^*M \times \mathbb{R}$ possesses a natural contact form and that contact flows are defined in order to preserve this form. Given a continuous function (contact Hamiltonian) $G : (x, p, u) \mapsto G(x, p, u)$ one can study viscosity solutions of stationary equations or evolutionary equations. We will mainly focus on the discounted approximation here but references on other topics were mentioned in the introduction and let us highlight also a recent beautiful paper that studies Lyapunov stability of the evolutionary semigroup in a general framework quite similar to the one we will describe below [187].

Our standing hypotheses are that $p \mapsto G(x, p, u)$ is convex and superlinear for all (x, u) and that $\partial_u G$ exists and is continuous in a neighborhood of $T^*M \times \{0\}$. The classical Tonelli Hamiltonian that we are interested in is $H(x, p) = G(x, p, 0)$ and we consider $\alpha(0)$ its critical constant.

For $\lambda > 0$ we focus on solutions to

$$G\big(x, D_x u, \lambda u(x)\big) = \alpha(0), \quad x \in M \qquad (\lambda\text{HJ})$$

and in the asymptotic behavior of possible solutions as $\lambda \to 0$.

An increasing character of G with respect to the last variable is a very convenient hypothesis. A completely decreasing case was studied for Hamiltonians of the form $G(x, p, u) = H(x, p) - u$ in [90] where the authors obtain partial results but the general discounted convergence remains open for this type of monotonicity.

A sort of degenerate increasing setting was considered in [78] generalizing the results presented in paragraph of Sect. 4.6.4. We denote by $L_G : TM \times \mathbb{R} \to \mathbb{R}$ the Lagrangian associated to G. Note that increasing properties of G translate to decreasing properties of L_G and that $\partial_u L_G$ is also well defined and continuous in a neighborhood of $T^*M \times \{0\}$. The main results of this paper can be summarized as follows:

Theorem *Assume that for all $(x, p) \in T^*M$, $u \mapsto H(x, p, u)$ is nondecreasing (or equivalently $u \mapsto L_G(x, v, u)$ nonincreasing for all $(x, v) \in TM$).*

Assume moreover that for all Mather measures μ of $H = G(\cdot, \cdot, 0)$ the non degeneracy inequality $\int_{TM} \partial_u L_G(x, v, 0)\, \mathrm{d}\mu(x, v) < 0$ holds. Then there exists $\lambda_0 > 0$ such that for all $0 < \lambda < \lambda_0$ the Eq. (λHJ) admits a unique viscosity *solution* $u_\lambda : M \to \mathbb{R}$. *Moreover the family* u_λ *uniformly converges as* $\lambda \to 0$ *to a weak KAM solution* u_0 *of* H.

A chalenge is to understand in non monotone cases what happens. The first thing is that there might not be uniqueness of solutions anymore. A simple but telling example is $G(x, p, u) = p^2 + \sin(u)$ defined on $\mathbb{T} \times \mathbb{R} \times \mathbb{R}$ for which if $\lambda > 0$, the constant functions equal to 0, to $\frac{\pi}{\lambda}$ and $-\frac{\pi}{\lambda}$ are solutions of (λHJ). Therefore in non monotone settings, one can expect diverging families of solutions as $\lambda \to 0$. In [97] we manage to show that the previous example is representative of all possible behaviors of families of solutions when dropping the global non decreasing hypothesis. We prove a trichotomy Theorem:

Theorem *Assume that for all Mather measures* μ *of* $H = G(\cdot, \cdot, 0)$ *the non degeneracy inequality* $\int_{TM} \partial_u L_G(x, v, 0)\, \mathrm{d}\mu(x, v) < 0$ *holds. Then there exist a weak KAM solution* u_0 *of* H *and functions* θ *and* ε *from* $(0, 1]$ *to* $\mathbb{R}$ *such that* $\theta(\lambda) \to +\infty$ *and* $\varepsilon(\lambda) \to 0$ *as* $\lambda \to 0$ *and such that if* $\lambda \in (0, 1]$ *and* u_λ *is a solution to* (λHJ) *then one of the three possibilities below hold:*

- $\|u_\lambda - u_0\|_\infty \leqslant \varepsilon(\lambda)$,
- $u_\lambda \leqslant -\theta(\lambda)$,
- $u_\lambda \geqslant \theta(\lambda)$.

Therefore, up to subsequences, solutions of the discounted equation either select a precise weak KAM solution of H, either diverge uniformly to $+\infty$ or to $-\infty$. The previous Theorem does not say anything about existence or uniqueness of solutions. Indeed, classical techniques of viscosity solutions theory fail without monotonicity. In the same article, we obtain results on those two problems for separable Hamiltonians of the form $G(x, p, u) = H(x, p) + \delta(x)u$ where $\delta : M \to \mathbb{R}$ is a continuous function. Note that the derivability property $\partial_u G$ is automatically satisfied. Recall also that in this case $L_G(x, v, u) = L(x, v) - \delta(x)u$ where L is the Lagrangian associated to H. The non degeneracy hypothesis then becomes $\int_{TM} \delta(x)\, \mathrm{d}\mu(x, v) > 0$ for all Mather measure μ of H. The discounted equation has a form we already encountered:

$$\lambda\delta(x)u(x) + H(x, D_x u) = \alpha(0), \quad x \in M \qquad (\lambda\delta\text{HJ})$$

Under those hypotheses we obtain the following existence (and convergence) result:

Theorem *Under the previous hypotheses there exists* $\lambda_0 > 0$ *such that for all* $\lambda \in (0, \lambda_0)$ *there exists a solution* $v_\lambda : M \to \mathbb{R}$ *of* ($\lambda\delta$HJ) *that is the (pointwise) maximal solution of* ($\lambda\delta$HJ). *Moreover the family* $(v_\lambda)_{\lambda \in (0,\lambda_0)}$ *converges as* $\lambda \to 0$.

Therefore, the previous Theorem excludes the possibility of families of solutions diverging to $+\infty$. By strengthening the hypothesis, we obtain other solutions (compare with Sect. 4.6.4):

Theorem *Assuming the previous hypotheses, suppose moreover that there exists $x_0 \in M$ such that $\delta(x_0) < 0$.*

1. *If there exists a neighborhood V of the projected Aubry set $\mathcal{A}_H$ such that $\delta_{|V} \geqslant 0$ then there exists a second family u_λ of solutions for $0 < \lambda < \lambda_0$ such that $u_\lambda \to -\infty$ as $\lambda \to 0$.*
2. *If $\delta_{|\mathcal{A}_H} > 0$ and if $(w_{\lambda_n})_n$ is any sequence of solutions with $\lambda_n \to 0$ and $w_{\lambda_n} \neq v_{\lambda_n}$ for all $n > 0$ then $w_{\lambda_n} \to -\infty$.*

The first point therefore gives a criterium for which there are at least two families of solutions, one converging and one diverging. The second point, asserts that under a slightly stronger hypothesis, the family $(v_\lambda)_{\lambda \in (0,\lambda_0)}$ of maximal solutions is the only family of solutions that converges as $\lambda \to 0$.

Weak KAM in Celestial Mechanics

Since at least Poincaré, celestial mechanics has been an amazing playground for mathematicians. Lack of compactness and collisions between masses complicate everything and make most classical technics in dynamical systems fail.

The setting is that of N punctual objects of masses $m_1, \cdots, m_N > 0$ evolving in the euclidean space $\mathbb{R}^n$. The position of the mass m_i is denoted by $r_i \in \mathbb{R}^n$ and its velocity $v_i \in \mathbb{R}^n$. We denote by $r_{ij} = \|r_i - r_j\|$ the euclidean distance between the masses m_i and m_j. We introduce as well the vectors $x = (r_1, \cdots, r_N)$ and $v = (v_1, \cdots, v_N)$. If $\kappa \in (0, 1)$ we introduce a potential

$$U_\kappa(x) = \sum_{i<j} m_i m_j r_{ij}^{-2\kappa}.$$

The classical Newtonian potential corresponds to the case $\kappa = \frac{1}{2}$. The Lagrangian function is

$$L(x, v) = \frac{1}{2}\sum_{i=1}^{N} m_i \|v_i\|^2 + U_\kappa(x).$$

With this at hand, one easily defines an action functional and a Lax–Oleinik semigroup by copy pasting the definitions for example the Lax–Oleinik is, for f a continuous function on $(\mathbb{R}^n)^N$, and $t > 0$

$$\forall x \in (\mathbb{R}^n)^N, \quad T^{-t} f(x) = \inf_\gamma f\big(\gamma(-t)\big) + \int_{-t}^{0} L\big(\gamma(s), \dot\gamma(s)\big)\, \mathrm{d}s,$$

where the infimum is taken on absolutely continuous curves such that $\gamma(0) = x$. The problem is that the Lagrangian is unbounded and even takes value $+\infty$ when two masses collide. Moreover the base manifold is not compact. Maderna manages to handle those difficulties and proves a weak KAM Theorem [155]:

Theorem *There exists a function* $u : (\mathbb{R}^n)^N \to \mathbb{R}$ *such that* $T^{-t}u = u$ *for all* $t > 0$.

Maderna also provides an equivariant version by proving the existence of such functions having the same symmetries as those of U_κ. The degeneracy of the potential translate to lesser regularity of the action functional and of weak KAM solutions. They are proven to be Hölder and not anymore locally Lipschitz.

Maderna pushed his study of weak KAM solutions for the N–body problem in collaboration with Venturelli. They obtained a striking and powerful existence result of trajectories having prescribed asymptotic behavior in [156].

Narrowing in on the Newtonian case $\kappa = \frac{1}{2}$, a trajectory $\big(r_1(t), \cdots, r_N(t)\big)$, defined for $t \geqslant t_0$ is called hyperbolic if for all $1 \leqslant i \leqslant N$ the velocity $\dot r_i(t) \to a_i$ converges as $t \to +\infty$ and all the limit vectors a_i are different. Let $\Omega \subset (\mathbb{R}^n)^N$ the set of such vectors without collisions: $x = (a_1, \cdots, a_N) \in \Omega$ if and only if $a_i \neq a_j$ for all $i \neq j$. The main Theorem of [156] is then:

Theorem *For the Newtonian N–body problem in* $\mathbb{R}^n$, *with* $n \geqslant 2$, *there are hyperbolic motions* $x : [0, +\infty) \to (\mathbb{R}^n)^N$ *such that*

$$x(t) = \sqrt{2h}ta + o(t),$$

as $t \to +\infty$. *This holds for any choice of* $x_0 = x(0) \in (\mathbb{R}^n)^N$, *for any limit speeds* $a \in \Omega$ *normalized such that* $\|a\| = 1$ *and any choice of constant* $h > 0$.

References

1. Al-Aidarous, E.S., Alzahrani, E. O., Ishii, H., Younas, A.M.M.: Asymptotic analysis for the eikonal equation with the dynamical boundary conditions. Math. Nachr. **287**(14–15), 1563–1588 (2014)
2. Albano, P., Cannarsa, P., Nguyen, K.T., Sinestrari, C.: Singular gradient flow of the distance function and homotopy equivalence. Math. Ann. **356**(1), 23–43 (2013)
3. Arcostanzo, M.: The C^0 integrability of symplectic twist maps without conjugate points. Ergodic Theory Dyn. Syst. **41**(1), 48–65 (2021)
4. Arcostanzo, M., Arnaud, M.C., Bolle, P., Zavidovique, M.: Tonelli Hamiltonians without conjugate points and C^0–integrability. Math. Z. **280**(1–2), 165–194 (2015)
5. Arnaud, M.C.: Fibrés de Green et régularité des graphes C^0-lagrangiens invariants par un flot de Tonelli. Ann. Henri Poincaré **9**(5), 881–926 (2008)
6. Arnaud, M.C.: Three results on the regularity of the curves that are invariant by an exact symplectic twist map. Publ. Math. Inst. Hautes Études Sci. **109**, 1–17 (2009)
7. Arnaud, M.C.: Green bundles and related topics. In: Proceedings of the International Congress of Mathematicians. Volume III, pp. 1653–1679. Hindustan Book Agency, New Delhi (2010)
8. Arnaud, M.C.: On a theorem due to Birkhoff. Geom. Funct. Anal. **20**(6), 1307–1316 (2010)
9. Arnaud, M.C.: The link between the shape of the irrational Aubry-Mather sets and their Lyapunov exponents. Ann. Math. **174**(3), 1571–1601 (2011)
10. Arnaud, M.C.: A nondifferentiable essential irrational invariant curve for a C^1 symplectic twist map. J. Mod. Dyn. **5**(3), 583–591 (2011)
11. Arnaud, M.C.: A particular minimization property implies C^0-integrability. J. Differ. Equ. **250**(5), 2389–2401 (2011)
12. Arnaud, M.C.: Pseudographs and the Lax-Oleinik semi-group: a geometric and dynamical interpretation. Nonlinearity **24**(1), 71–78 (2011)
13. Arnaud, M.C.: Green bundles, Lyapunov exponents and regularity along the supports of the minimizing measures. Ann. Inst. H. Poincaré Anal. Non Linéaire **29**(6), 989–1007 (2012)
14. Arnaud, M.C.: Lower and upper bounds for the Lyapunov exponents of twisting dynamics: a relationship between the exponents and the angle of Oseledets' splitting. Ergodic Theory Dyn. Syst. **33**(3), 693–712 (2013)
15. Arnaud, M.C.: Boundaries of instability zones for symplectic twist maps. J. Inst. Math. Jussieu **13**(1), 19–41 (2014)
16. Arnaud, M.C.: Hyperbolicity for conservative twist maps of the 2-dimensional annulus. Publ. Mat. Urug. **16**, 1–39 (2016)
17. Arnaud, M.C.: Lyapunov exponents for conservative twisting dynamics: a survey. In: Ergodic Theory, pp. 108–133. De Gruyter, Berlin (2016)

M. Zavidovique, *Discrete Weak KAM Theory*, Lecture Notes in Mathematics 2377,
https://doi.org/10.1007/978-3-031-96809-9

18. Arnaud, M.C.: Lyapunov exponents of minimizing measures for globally positive diffeomorphisms in all dimensions. Commun. Math. Phys. **343**(3), 783–810 (2016)
19. Arnaud, M.C., Su, X.: On the C^1 and C^2-convergence to weak K.A.M. solutions. Commun. Math. Phys. **392**(3), 825–861 (2022)
20. Arnaud, M.C., Venturelli, A.: A multidimensional Birkhoff theorem for time-dependent Tonelli Hamiltonians. Calc. Var. Partial Differ. Equ. **56**(4), paper No. 122 27 (2017)
21. Arnaud, M.C., Zavidovique, M.: On the transversal dependence of weak K.A.M. solutions for symplectic twist maps (2018). arXiv1809.02372
22. Arnaud, M.C., Zavidovique, M.: Actions of symplectic homeomorphisms/diffeomorphisms on foliations by curves in dimension 2. Ergodic Theory Dyn. Syst. **43**(3), 794–826 (2023)
23. Arnaud, M.C., Zavidovique, M.: Weak K.A.M. solutions and minimizing orbits of twist maps. Trans. Am. Math. Soc. **376**(11), 8129–8171 (2023)
24. Arnol'd, V.I.: Proof of a theorem of A. N. Kolmogorov on the preservation of conditionally periodic motions under a small perturbation of the Hamiltonian. Uspehi Mat. Nauk **18**(5), 13–40 (1963)
25. Arnol'd, V.I.: Small denominators and problems of stability of motion in classical and celestial mechanics. Uspehi Mat. Nauk **18**(6), 91–192 (1963)
26. Arnol'd, V.I.: Instability of dynamical systems with many degrees of freedom. Dokl. Akad. Nauk SSSR **156**, 9–12 (1964)
27. Aubry, S., Le Daeron, P.Y.: The discrete Frenkel-Kontorova model and its extensions. I. Exact results for the ground-states. Phys. D **8**(3), 381–422 (1983)
28. Audin, M., Damian, M.: Théorie de Morse et homologie de Floer. Savoirs Actuels (Les Ulis). [Current Scholarship (Les Ulis)]. EDP Sciences, Les Ulis; CNRS Éditions, Paris (2010)
29. Avila, A., Fayad, B.: Non-differentiable irrational curves for C^1 twist map. Ergodic Theory Dyn. Syst. **42**(2), 491–499 (2022)
30. Bangert, V.: Mather sets for twist maps and geodesics on tori. In: Kirchgraber, U., Walther, H.O. (eds.) Dynamics Reported, pp. 1–56. Vieweg+Teubner Verlag, Wiesbaden (1988)
31. Bangert, V.: Geodesic rays, Busemann functions and monotone twist maps. Calc. Var. Partial Differ. Equ. **2**(1), 49–63 (1994)
32. Bardi, M., Capuzzo-Dolcetta, I.: Optimal Control and Viscosity Solutions of Hamilton-Jacobi-Bellman Equations. Systems & Control: Foundations & Applications. Birkhäuser, Boston (1997). With appendices by Maurizio Falcone and Pierpaolo Soravia
33. Barles, G.: Solutions de viscosité des équations de Hamilton-Jacobi. Mathématiques & Applications (Berlin) [Mathematics & Applications], vol. 17. Springer, Paris (1994)
34. Barles, G., Souganidis, P.E.: Convergence of approximation schemes for fully nonlinear second order equations. Asymptotic Anal. **4**(3), 271–283 (1991)
35. Barles, G., Souganidis, P.E.: On the large time behavior of solutions of Hamilton-Jacobi equations. SIAM J. Math. Anal. **31**(4), 925–939 (2000)
36. Barles, G., Souganidis, P.E.: Some counterexamples on the asymptotic behavior of the solutions of Hamilton-Jacobi equations. C. R. Acad. Sci. Paris Sér. I Math. **330**(11), 963–968 (2000)
37. Barron, E.N., Jensen, R.: Semicontinuous viscosity solutions for Hamilton-Jacobi equations with convex Hamiltonians. Commun. Partial Differ Equ. **15**(12), 1713–1742 (1990)
38. Barron, E.N., Jensen, R.: Optimal control and semicontinuous viscosity solutions. Proc. Am. Math. Soc. **113**(2), 397–402 (1991)
39. Berger, P., Turaev, D.: On Herman's positive entropy conjecture. Adv. Math. **349**, 1234–1288 (2019)
40. Berger, P., Yoccoz, J.C.: Strong regularity. Société Mathématique de France, Paris (2019). Astérisque No. 410 (2019)
41. Bernard, P.: Connecting orbits of time dependent Lagrangian systems. Ann. Inst. Fourier **52**(5), 1533–1568 (2002)
42. Bernard, P.: Existence of $C^{1,1}$ critical sub-solutions of the Hamilton-Jacobi equation on compact manifolds. Ann. Sci. École Norm. Sup. **40**(3), 445–452 (2007)

43. Bernard, P.: The dynamics of pseudographs in convex Hamiltonian systems. J. Am. Math. Soc. **21**(3), 615–669 (2008)
44. Bernard, P.: Large normally hyperbolic cylinders in a priori stable Hamiltonian systems. Ann. Henri Poincaré **11**(5), 929–942 (2010)
45. Bernard, P.: Lasry-Lions regularization and a lemma of Ilmanen. Rend. Semin. Mat. Univ. Padova **124**, 221–229 (2010)
46. Bernard, P.: On the number of Mather measures of Lagrangian systems. Arch. Ration. Mech. Anal. **197**(3), 1011–1031 (2010)
47. Bernard, P., Buffoni, B.: The Monge problem for supercritical Mañé potentials on compact manifolds. Adv. Math. **207**(2), 691–706 (2006)
48. Bernard, P., Buffoni, B.: Optimal mass transportation and Mather theory. J. Eur. Math. Soc. **9**(1), 85–121 (2007)
49. Bernard, P., Buffoni, B.: Weak KAM pairs and Monge-Kantorovich duality. In: Asymptotic Analysis and Singularities—Elliptic and Parabolic PDEs and Related Problems. Advanced Studies in Pure Mathematics, vol. 47, pp. 397–420. The Mathematical Society of Japan, Tokyo (2007)
50. Bernard, P., Contreras, G.: A generic property of families of Lagrangian systems. Ann. Math. **167**(3), 1099–1108 (2008)
51. Bernard, P., Roquejoffre, J.M.: Convergence to time-periodic solutions in time-periodic Hamilton-Jacobi equations on the circle. Commun. Partial Differ. Equ. **29**(3–4), 457–469 (2004)
52. Bernard, P., Suhr, S.: Lyapounov functions of closed cone fields: from Conley theory to time functions. Commun. Math. Phys. **359**(2), 467–498 (2018)
53. Bernard, P., Suhr, S.: Cauchy and uniform temporal functions of globally hyperbolic cone fields. Proc. Am. Math. Soc. **148**(11), 4951–4966 (2020)
54. Bernard, P., Zavidovique, M.: Regularization of subsolutions in discrete weak KAM theory. Can. J. Math. **65**(4), 740–756 (2013)
55. Bernard, P., Kaloshin, V., Zhang, K.: Arnold diffusion in arbitrary degrees of freedom and normally hyperbolic invariant cylinders. Acta Math. **217**(1), 1–79 (2016)
56. Bernardi, O., Florio, A.: A Conley-type decomposition of the strong chain recurrent set. Ergodic Theory Dyn. Syst. **39**(5), 1261–1274 (2019)
57. Bernardi, O., Florio, A.: Existence of Lipschitz continuous Lyapunov functions strict outside the strong chain recurrent set. Dyn. Syst. **34**(1), 71–92 (2019)
58. Bernardi, O., Florio, A., Wiseman, J.: The generalized recurrent set, explosions and Lyapunov functions. J. Dyn. Differ. Equ. **32**(4), 1797–1817 (2020)
59. Birkhoff, G.D.: Sur quelques courbes fermées remarquables. Bull. Soc. Math. France **60**, 1–26 (1932)
60. Bouillard, A., Faou, E., Zavidovique, M.: Fast weak-KAM integrators for separable Hamiltonian systems. Math. Comput. **85**(297), 85–117 (2016)
61. Burago, D., Ivanov, S.: Riemannian tori without conjugate points are flat. Geom. Funct. Anal. **4**(3), 259–269 (1994)
62. Cagnetti, F., Gomes, D., Mitake, H., Tran, H.V.: A new method for large time behavior of degenerate viscous Hamilton-Jacobi equations with convex Hamiltonians. Ann. Inst. H. Poincaré Anal. Non Linéaire **32**(1), 183–200 (2015)
63. Camilli, F., Ley, O., Loreti, P.: Homogenization of monotone systems of Hamilton-Jacobi equations. ESAIM Control Optim. Calc. Var. **16**(1), 58–76 (2010)
64. Camilli, F., Ley, O., Loreti, P., Nguyen, V.D.: Large time behavior of weakly coupled systems of first-order Hamilton-Jacobi equations. NoDEA Nonlinear Differ. Equ. Appl. **19**(6), 719–749 (2012)
65. Cannarsa, P., Cheng, W.: Generalized characteristics and Lax-Oleinik operators: global theory. Calc. Var. Partial Differ. Equ. **56**(5), paper No. 125, 31 (2017)
66. Cannarsa, P., Cheng, W.: Singularities of solutions of Hamilton-Jacobi equations. Milan J. Math. **89**(1), 187–215 (2021)

67. Cannarsa, P., Sinestrari, C.: Semiconcave Functions, Hamilton-Jacobi Equations, and Optimal Control. Progress in Nonlinear Differential Equations and their Applications, vol. 58. Birkhäuser, Boston (2004)
68. Cannarsa, P., Yu, Y.: Singular dynamics for semiconcave functions. J. Eur. Math. Soc. **11**(5), 999–1024 (2009)
69. Cannarsa, P., Mazzola, M., Sinestrari, C.: Global propagation of singularities for time dependent Hamilton-Jacobi equations. Discrete Contin. Dyn. Syst. **35**(9), 4225–4239 (2015)
70. Cannarsa, P., Cheng, W., Fathi, A.: On the topology of the set of singularities of a solution to the Hamilton-Jacobi equation. C. R. Math. Acad. Sci. Paris **355**(2), 176–180 (2017)
71. Cannarsa, P., Cheng, W., Jin, L., Wang, K., Yan, J.: Herglotz' variational principle and Lax-Oleinik evolution. J. Math. Pures Appl. **141**, 99–136 (2020)
72. Cannarsa, P., Cheng, W., Fathi, A.: Singularities of solutions of time dependent Hamilton-Jacobi equations. Applications to Riemannian geometry. Publ. Math. Inst. Hautes Études Sci. **133**, 327–366 (2021)
73. Cardaliaguet, P., Porretta, A.: Long time behavior of the master equation in mean field game theory. Anal. PDE **12**(6), 1397–1453 (2019)
74. Carneiro, M.J.D.: On minimizing measures of the action of autonomous Lagrangians. Nonlinearity **8**(6), 1077–1085 (1995)
75. Charfi, S.: Diverging solutions of time periodic Hamilton–Jacobi equations for time periodic Tonelli Hamiltonians. Personal Communication (2025)
76. Chen, Q.: Convergence of solutions of Hamilton-Jacobi equations depending nonlinearly on the unknown function. Adv. Calculus Var. **16**(1), 45–68 (2023)
77. Chen, Q., Cheng, W., Ishii, H., Zhao, K.: Vanishing contact structure problem and convergence of the viscosity solutions. Commun. Partial Differ. Equ. **44**(9), 801–836 (2019)
78. Chen, Q., Fathi, A., Zavidovique, M., Zhang, J.: Convergence of the solutions of the nonlinear discounted Hamilton-Jacobi equation: the central role of Mather measures. J. Math. Pures Appl. **181**, 22–57 (2024)
79. Cheng, C.Q., Xue, J.: Order property and modulus of continuity of weak KAM solutions. Calc. Var. Partial Differ. Equ. **57**(2), paper no. 65, 27 (2018)
80. Cheng, J., Sun, Y.: A necessary and sufficient condition for a twist map being integrable. Sci. China Ser. A **39**(7), 709–717 (1996)
81. Clarke, F. (2013). Functional Analysis, Calculus of Variations and Optimal Control. Graduate Texts in Mathematics, vol. 264. Springer, London (2013)
82. Contreras, G.: Ground states are generically a periodic orbit. Invent. Math. **205**(2), 383–412 (2016)
83. Contreras, G.: Proof of the C^2 Mañé's conjecture on surfaces (2024). https://arxiv.org/abs/2408.01009
84. Contreras, G., Figalli, A., Rifford, L.: Generic hyperbolicity of Aubry sets on surfaces. Invent. Math. **200**(1), 201–261 (2015)
85. Contreras, G., Iturriaga, R., Siconolfi, A.: Homogenization on arbitrary manifolds. Calc. Var. Partial Differ. Equ. **52**(1–2), 237–252 (2015)
86. Crandall, M.G., Lions, P.L.: Viscosity solutions of Hamilton-Jacobi equations. Trans. Am. Math. Soc. **277**(1), 1–42 (1983)
87. Crandall, M.G., Ishii, H., Lions, P.L.: Uniqueness of viscosity solutions of Hamilton-Jacobi equations revisited. J. Math. Soc. Japan **39**(4), 581–596 (1987)
88. Crandall, M.G., Ishii, H., Lions, P.L.: User's guide to viscosity solutions of second order partial differential equations. Bull. Am. Math. Soc. **27**(1), 1–67 (1992)
89. Davini, A., Siconolfi, A.: A generalized dynamical approach to the large time behavior of solutions of Hamilton-Jacobi equations. SIAM J. Math. Anal. **38**(2), 478–502 (2006)
90. Davini, A., Wang, L.: On the vanishing discount problem from the negative direction. Discrete Contin. Dyn. Syst. A **41**(5), 2377 (2021)
91. Davini, A., Zavidovique, M.: Aubry sets for weakly coupled systems of Hamilton-Jacobi equations. SIAM J. Math. Anal. **46**(5), 3361–3389 (2014)

92. Davini, A., Zavidovique, M.: On the (non) existence of viscosity solutions of multi-time Hamilton-Jacobi equations. J. Differ. Equ. **258**(2), 362–378 (2015)
93. Davini, A., Zavidovique, M.: Convergence of the solutions of discounted Hamilton-Jacobi systems. Adv. Calc. Var. **14**(2), 193–206 (2021)
94. Davini, A., Fathi, A., Iturriaga, R., Zavidovique, M.: Convergence of the solutions of the discounted equation: the discrete case. Math. Z. **284**(3–4), 1021–1034 (2016)
95. Davini, A., Fathi, A., Iturriaga, R., Zavidovique, M.: Convergence of the solutions of the discounted Hamilton-Jacobi equation: convergence of the discounted solutions. Invent. Math. **206**(1), 29–55 (2016)
96. Davini, A., Siconolfi, A., Zavidovique, M.: Random Lax-Oleinik semigroups for Hamilton-Jacobi systems. J. Math. Pures Appl. **120**, 294–333 (2018)
97. Davini, A., Ni, P., Yan, J., Zavidovique, M.: Convergence/divergence phenomena in the vanishing discount limit of Hamilton-Jacobi equations (2024). https://arxiv.org/abs/2411.13780
98. Du, J., Su, X.: Singular dynamics for discrete weak K.A.M. solutions of exact twist maps (2024). https://arxiv.org/abs/2403.01141
99. Dugundji, J.: Topology. Allyn and Bacon, Boston (1966)
100. Duistermaat, J.J.: On global action-angle coordinates. Commun. Pure Appl. Math. **33**(6), 687–706 (1980)
101. Evans, L.C.: The perturbed test function method for viscosity solutions of nonlinear PDE. Proc. Roy. Soc. Edinburgh Sect. A **111**(3–4), 359–375 (1989)
102. Evans, L.C.: Adjoint and compensated compactness methods for Hamilton-Jacobi PDE. Arch. Ration. Mech. Anal. **197**(3), 1053–1088 (2010)
103. Evans, L.C.: Partial Differential Equations. American Mathematical Society, Providence (2010)
104. Fathi, A.: Solutions KAM faibles conjuguées et barrières de Peierls. C. R. Acad. Sci. Paris Sér. I Math. **325**(6), 649–652 (1997)
105. Fathi, A.: Théorème KAM faible et théorie de Mather sur les systèmes lagrangiens. C. R. Acad. Sci. Paris Sér. I Math. **324**(9), 1043–1046 (1997)
106. Fathi, A.: Orbites hétéroclines et ensemble de Peierls. C. R. Acad. Sci. Paris Sér. I Math. **326**(10), 1213–1216 (1998)
107. Fathi, A.: Sur la convergence du semi-groupe de Lax-Oleinik. C. R. Acad. Sci. Paris Sér. I Math. **327**(3), 267–270 (1998)
108. Fathi, A.: Regularity of C^1 solutions of the Hamilton-Jacobi equation. Ann. Fac. Sci. Toulouse Math. **12**(4), 479–516 (2003)
109. Fathi, A.: Weak KAM Theorem in Lagrangian Dynamics, preliminary version 10, Lyon. Unpublished (2008)
110. Fathi, A., Figalli, A.: Optimal transportation on non-compact manifolds. Isr. J. Math. **175**, 1–59 (2010)
111. Fathi, A., Mather, J.N.: Failure of convergence of the Lax-Oleinik semi-group in the time-periodic case. Bull. Soc. Math. France **128**(3), 473–483 (2000)
112. Fathi, A., Pageault, P.: Aubry-Mather theory for homeomorphisms. Ergodic Theory Dyn. Syst. **35**(4), 1187–1207 (2015)
113. Fathi, A., Pageault, P.: Smoothing Lyapunov functions. Trans. Am. Math. Soc. **371**(3), 1677–1700 (2019)
114. Fathi, A., Siconolfi, A.: Existence of C^1 critical subsolutions of the Hamilton-Jacobi equation. Invent. Math. **155**(2), 363–388 (2004)
115. Fathi, A., Siconolfi, A.: PDE aspects of Aubry-Mather theory for quasiconvex Hamiltonians. Calc. Var. Partial Differ. Equ. **22**(2), 185–228 (2005)
116. Fathi, A., Siconolfi, A.: On smooth time functions. Math. Proc. Cambridge Philos. Soc. **152**(2), 303–339 (2012)
117. Fathi, A., Zavidovique, M.: Ilmanen's lemma on insertion of $C^{1,1}$ functions. Rend. Semin. Mat. Univ. Padova **124**, 203–219 (2010)

118. Fathi, A., Figalli, A., Rifford, L.: On the Hausdorff dimension of the Mather quotient. Commun. Pure Appl. Math. **62**(4), 445–500 (2009)
119. Féjoz, J.: Démonstration du ‘théorème d’Arnold’ sur la stabilité du système planétaire (d’après Herman). Ergodic Theory Dyn. Syst. **24**(5), 1521–1582 (2004)
120. Féjoz, J.: A proof of the invariant torus theorem of Kolmogorov. Regul. Chaotic Dyn. **17**(1), 1–5 (2012)
121. Figalli, A., Rifford, L.: Closing Aubry sets I. Commun. Pure Appl. Math. **68**(2), 210–285 (2015)
122. Figalli, A., Rifford, L.: Closing Aubry sets II. Commun. Pure Appl. Math. **68**(3), 345–412 (2015)
123. Garibaldi, E., Thieullen, P.: An ergodic description of ground states. J. Stat. Phys. **158**(2), 359–371 (2015)
124. Garibaldi, E., Lopes, A.O., Thieullen, P.: On calibrated and separating sub-actions. Bull. Braz. Math. Soc. **40**(4), 577–602 (2009)
125. Goebel, K., Kirk, W.A.: Topics in Metric Fixed Point Theory. Cambridge Studies in Advanced Mathematics, vol. 28. Cambridge University Press, Cambridge (1990)
126. Golé, C.: Symplectic Twist Maps. Advanced Series in Nonlinear Dynamics, vol. 18. World Scientific, River Edge (2001). Global variational techniques
127. Gomes, D.A.: Generalized Mather problem and selection principles for viscosity solutions and Mather measures. Adv. Calc. Var. **1**(3), 291–307 (2008)
128. Heber, J.: On the geodesic flow of tori without conjugate points. Math. Z. **216**(2), 209–216 (1994)
129. Herman, M.R.: Sur les courbes invariantes par les difféomorphismes de l’anneau. Vol. 1. Astérisque, vol. 103. Société Mathématique de France, Paris (1983). With an appendix by Albert Fathi, With an English summary
130. Herman, M.R.: Sur les courbes invariantes par les difféomorphismes de l’anneau. Vol. 2. Astérisque **144**, 248 (1986). With a correction to: ıt On the curves invariant under diffeomorphisms of the annulus, Vol. 1 (French) [Astérisque No. 103-104, Soc. Math. France, Paris, 1983; MR0728564 (85m:58062)]
131. Ibrahim, H., Siconolfi, A., Zabad, S.: Cycle characterization of the Aubry set for weakly coupled Hamilton-Jacobi systems. Commun. Contemp. Math. **20**(6), 1750095, 28 (2018)
132. Ilmanen, T.: The level-set flow on a manifold. In: Differential Geometry: Partial Differential Equations on Manifolds (Los Angeles, CA, 1990). Proceedings of Symposia in Pure Mathematics, vol. 54, pp. 193–204. American Mathematical Society, Providence (1993)
133. Ishii, H.: Perron’s method for Hamilton-Jacobi equations. Duke Math. J. **55**(2), 369–384 (1987)
134. Ishii, H.: On uniqueness and existence of viscosity solutions of fully nonlinear second-order elliptic PDEs. Commun. Pure Appl. Math. **42**(1), 15–45 (1989)
135. Ishii, H.: Asymptotic solutions for large time of Hamilton-Jacobi equations in Euclidean n space. Ann. Inst. H. Poincaré Anal. Non Linéaire **25**(2), 231–266 (2008)
136. Ishii, H.: The vanishing discount problem for monotone systems of Hamilton-Jacobi equations. Part 1: linear coupling. Math. Eng. **3**(4), paper no. 032, 21 (2021)
137. Ishii, H.: The vanishing discount problem for monotone systems of Hamilton-Jacobi equations: a counterexample to the full convergence. Math. Eng. **5**(4), paper no. 072, 10 (2023)
138. Ishii, H., Jin, L.: The vanishing discount problem for monotone systems of Hamilton-Jacobi equations: part 2—nonlinear coupling. Calc. Var. Partial Differ. Equ. **59**(4), paper no. 140, 28 (2020)
139. Ishii, H., Lions, PL.: Viscosity solutions of fully nonlinear second-order elliptic partial differential equations. J. Differ. Equ. **83**(1), 26–78 (1990)
140. Ishii, H., Siconolfi, A.: The vanishing discount problem for Hamilton-Jacobi equations in the Euclidean space. Commun. Partial Differ. Equ. **45**(6), 525–560 (2020)
141. Ishii, H., Mitake, H., Tran, H.V.: The vanishing discount problem and viscosity Mather measures. Part 1: the problem on a torus. J. Math. Pures Appl. **108**(2), 125–149 (2017)

142. Ishii, H., Mitake, H., Tran, H.V.: The vanishing discount problem and viscosity Mather measures. Part 2: boundary value problems. J. Math. Pures Appl. **108**(3), 261–305 (2017)
143. Iturriaga, R., Sánchez-Morgado, H.: Limit of the infinite horizon discounted Hamilton-Jacobi equation. Discrete Contin. Dyn. Syst. Ser. B **15**(3), 623–635 (2011)
144. Kaloshin, V., Zhang, K.: Arnold Diffusion for Smooth Systems of Two and a Half Degrees of Freedom. Annals of Mathematics Studies, vol. 208. Princeton University Press, Princeton (2020)
145. Kantorovitch, L.: On the translocation of masses. C. R. (Doklady) Acad. Sci. URSS (N.S.) **37**, 199–201 (1942)
146. Katznelson, Y., Ornstein, D.S.: Twist maps and Aubry-Mather sets. In: Lipa's Legacy (New York, 1995). Contemporary Mathematics, vol. 211, pp. 343–357. American Mathematical Society, Providence (1997)
147. Kirk, W.A., Sims, B. (Eds.): Handbook of Metric Fixed Point Theory. Kluwer Academic Publishers, Dordrecht (2001)
148. Kolmogorov, A.N.: On conservation of conditionally periodic motions for a small change in Hamilton's function. Dokl. Akad. Nauk SSSR (N.S.) **98**, 527–530 (1954)
149. Lasry, J.M., Lions, P.L.: A remark on regularization in Hilbert spaces. Isr. J. Math. **55**(3), 257–266 (1986)
150. Le Calvez, P.: Propriétés des attracteurs de Birkhoff. Ergodic Theory Dyn. Syst. **8**(2), 241–310 (1988)
151. Lions, P.L.: Generalized Solutions of Hamilton-Jacobi Equations. Research Notes in Mathematics, vol. 69. Pitman (Advanced Publishing Program), Boston (1982)
152. Lions, P.L., Papanicolaou, G., Varadhan, S.R.S.: Homogenization of Hamilton-Jacobi equation. Unpublished preprint (1987)
153. Mañé, R.: On the minimizing measures of Lagrangian dynamical systems. Nonlinearity **5**(3), 623–638 (1992)
154. Mañé, R.: Generic properties and problems of minimizing measures of Lagrangian systems. Nonlinearity **9**(2), 273–310 (1996)
155. Maderna, E.: On weak KAM theory for N-body problems. Ergodic Theory Dyn. Syst. **32**(3), 1019–1041 (2012)
156. Maderna, E., Venturelli, A.: Viscosity solutions and hyperbolic motions: a new PDE method for the N-body problem. Ann. Math. **192**(2), 499–550 (2020)
157. Massart, D.: On Aubry sets and Mather's action functional. Isr. J. Math. **134**, 157–171 (2003)
158. Massart, D., Sorrentino, A.: Differentiability of Mather's average action and integrability on closed surfaces. Nonlinearity **24**(6), 1777–1793 (2011)
159. Mather, J.N.: Existence of quasiperiodic orbits for twist homeomorphisms of the annulus. Topology **21**(4), 457–467 (1982)
160. Mather, J.N.: Nonexistence of invariant circles. Ergodic Theory Dyn. Syst. **4**(2), 301–309 (1984)
161. Mather, J.N.: A criterion for the nonexistence of invariant circles. Inst. Hautes Études Sci. Publ. Math. **63**, 153–204 (1986)
162. Mather, J.N.: Modulus of continuity for Peierls's barrier. In: *Periodic Solutions of Hamiltonian Systems and Related Topics (Il Ciocco, 1986)*. NATO ASI Series, Series C: Mathematical and Physical Sciences, vol. 209, pp. 177–202. Reidel, Dordrecht (1987)
163. Mather, J.N.: Minimal measures. Comment. Math. Helv. **64**(3), 375–394 (1989)
164. Mather, J.N.: Differentiability of the minimal average action as a function of the rotation number. Bol. Soc. Brasil. Mat. (N.S.) **21**(1), 59–70 (1990)
165. Mather, J.N.: Action minimizing invariant measures for positive definite Lagrangian systems. Math. Z. **207**(2), 169–207 (1991)
166. Mather, J.N.: Variational construction of orbits of twist diffeomorphisms. J. Am. Math. Soc. **4**(2), 207–263 (1991)
167. Mather, J.N.: Variational construction of connecting orbits. Ann. Inst. Fourier (Grenoble) **43**(5), 1349–1386 (1993)

168. Mather, J.N.: A property of compact, connected, laminated subsets of manifolds. Ergodic Theory Dyn. Syst. **22**(5), 1507–1520 (2002)
169. Mather, J.N.: Arnol' d diffusion. I. Announcement of results. Sovrem. Mat. Fundam. Napravl. **2**, 116–130 (2003)
170. Mather, J.N.: Total disconnectedness of the quotient Aubry set in low dimensions. Commun. Pure Appl. Math. **56**(8), 1178–1183 (2003). Dedicated to the memory of Jürgen K. Moser
171. Mather, J.N.: Arnold diffusion by variational methods. In: Essays in Mathematics and its Applications, pp. 271–285. Springer, Heidelberg (2012)
172. Mather, J.N., Forni, G.: Action minimizing orbits in Hamiltonian systems. In: Transition to Chaos in Classical and Quantum Mechanics (Montecatini Terme, 1991). Lecture Notes in Mathematics, pp. 92–186. Springer, Berlin (1994)
173. Mitake, H., Tran, H.V.: Selection problems for a discount degenerate viscous Hamilton-Jacobi equation. Adv. Math. **306**, 684–703 (2017)
174. Mitake, H., Siconolfi, A., Tran, H.V., Yamada, N.: A Lagrangian approach to weakly coupled Hamilton-Jacobi systems. SIAM J. Math. Anal. **48**(2), 821–846 (2016)
175. Monge, G.: Mémoire sur la théorie des déblais et des remblais. De l'Imprimerie Royale (1781)
176. Monzner, A., Vichery, N., Zapolsky, F.: Partial quasimorphisms and quasistates on cotangent bundles, and symplectic homogenization. J. Mod. Dyn. **6**(2), 205–249 (2012)
177. Moser, J.: On invariant curves of area-preserving mappings of an annulus. Nachr. Akad. Wiss. Göttingen Math.-Phys. Kl. II **1962**, 1–20 (1962)
178. Moser, J.: Monotone twist mappings and the calculus of variations. Ergodic Theory Dyn. Syst. **6**(3), 401–413 (1986)
179. Namah, G., Roquejoffre, J.M.: Comportement asymptotique des solutions d'une classe d'équations paraboliques et de Hamilton-Jacobi. C. R. Acad. Sci. Paris Sér. I Math. **324**(12), 1367–1370 (1997)
180. Ni, P., Zavidovique, M.: Nonlinear and degenerate discounted approximation in discrete weak KAM theory. Math. Z. **310**(4), paper no. 63, 27 (2025)
181. Pageault, P.: Conley barriers and their applications: chain-recurrence and Lyapunov functions. Topol. Appl. **156**(15), 2426–2442 (2009)
182. Phelps, R.R.: Lectures on Choquet's Theorem. Lecture Notes in Mathematics, vol. 1757, 2nd edn. Springer, Berlin (2001)
183. Pozza, M., Siconolfi, A.: Discounted Hamilton-Jacobi equations on networks and asymptotic analysis. Ind. Univ. Math. J. **70**(3), 1103–1129 (2021)
184. Reich, S.: Strong convergence theorems for resolvents of accretive operators in Banach spaces. J. Math. Anal. Appl. **75**(1), 287–292 (1980)
185. Rockafellar, R.T.: Convex Analysis. Princeton Landmarks in Mathematics. Princeton University Press, Princeton (1997). Reprint of the 1970 original, Princeton Paperbacks
186. Roos, V.: Variational and viscosity operators for the evolutionary Hamilton-Jacobi equation. Commun. Contemp. Math. **21**(4), 1850018, 76 (2019)
187. Ruan, Y., Wang, K., Yan, J.: Lyapunov stability and uniqueness problems for Hamilton-Jacobi equations without monotonicity. Comm. Math. Phys. **406**(7), paper no. 157, 50 (2025)
188. Siconolfi, A., Zabad, S.: Scalar reduction techniques for weakly coupled Hamilton-Jacobi systems. NoDEA Nonlinear Differ. Equ. Appl. **25**(6), paper no. 50, 20 (2018)
189. Sorrentino, A.: On the total disconnectedness of the quotient Aubry set. Ergodic Theory Dyn. Syst. **28**(1), 267–290 (2008)
190. Sorrentino, A.: Action-Minimizing Methods in Hamiltonian Dynamics. Mathematical Notes, vol. 50. Princeton University Press, Princeton (2015). An introduction to Aubry-Mather theory
191. Sorrentino, A.: On the homogenization of the Hamilton-Jacobi equation (2019). https://arxiv.org/abs/1904.01359
192. Souganidis, P.E.: Approximation schemes for viscosity solutions of Hamilton-Jacobi equations. J. Differ. Equ. **59**(1), 1–43 (1985)
193. Suhr, S.: Aubry-Mather theory for Lorentzian manifolds. J. Fixed Point Theory Appl. **21**(2), paper no. 71, 42 (2019)

194. Viterbo, C.: Symplectic homogenization. J. Éc. Polytech. Math. **10**, 67–140 (2023)
195. Walsh, J.A.: The dynamics of circle homeomorphisms: a hands-on introduction. Math. Mag. **72**(1), 3–13 (1999)
196. Wang, K., Wang, L., Yan, J.: Implicit variational principle for contact Hamiltonian systems. Nonlinearity **30**(2), 492–515 (2017)
197. Wang, K., Wang, L., Yan, J.: Aubry-Mather theory for contact Hamiltonian systems. Commun. Math. Phys. **366**(3), 981–1023 (2019)
198. Wang, K., Wang, L., Yan, J.: Variational principle for contact Hamiltonian systems and its applications. J. Math. Pures Appl. **123**, 167–200 (2019)
199. Wang, Y.N., Yan, J., Zhang, J.: Convergence of viscosity solutions of generalized contact Hamilton-Jacobi equations. Arch. Ration. Mech. Anal. **241**(2), 885–902 (2021)
200. Wei, Q.: Viscosity solution of the Hamilton-Jacobi equation by a limiting minimax method. Nonlinearity **27**(1), 17–41 (2014)
201. Zavidovique, M.: Existence of $C^{1,1}$ critical subsolutions in discrete weak KAM theory. J. Mod. Dyn. **4**(4), 693–714 (2010)
202. Zavidovique, M.: Strict sub-solutions and Mañé potential in discrete weak KAM theory. Comment. Math. Helv. **87**(1), 1–39 (2012)
203. Zavidovique, M.: Fixed points of contractions approximating 1-Lipschitz maps. Grad. J. Math. **4**(2), 56–61 (2019)
204. Zavidovique, M.: Twisted Lax-Oleinik formulas and weakly coupled systems of Hamilton-Jacobi equations. Ann. Fac. Sci. Toulouse Math. **28**(2), 209–224 (2019)
205. Zavidovique, M.: Convergence of solutions for some degenerate discounted Hamilton-Jacobi equations. Anal. PDE **15**(5), 1287–1311 (2022)
206. Zhang, J.: Global behaviors of weak KAM solutions for exact symplectic twist maps. J. Differ. Equ. **269**(7), 5730–5753 (2020)
207. Zhao, K., Cheng, W.: On the vanishing contact structure for viscosity solutions of contact type Hamilton-Jacobi equations I: Cauchy problem. Discrete Contin. Dyn. Syst. **39**(8), 4345–4358 (2019)
208. Ziliotto, B.: Convergence of the solutions of the discounted Hamilton-Jacobi equation: a counterexample. J. Math. Pures Appl. **128**, 330–338 (2019)

Index

M. Zavidovique, *Discrete Weak KAM Theory*, Lecture Notes in Mathematics 2377,
https://doi.org/10.1007/978-3-031-96809-9

LECTURE NOTES IN MATHEMATICS Springer

Editors in Chief: J.-M. Morel, B. Teissier;

Editorial Policy

1. Lecture Notes aim to report new developments in all areas of mathematics and their applications – quickly, informally and at a high level. Mathematical texts analysing new developments in modelling and numerical simulation are welcome.

   Manuscripts should be reasonably self-contained and rounded off. Thus they may, and often will, present not only results of the author but also related work by other people. They may be based on specialised lecture courses. Furthermore, the manuscripts should provide sufficient motivation, examples and applications. This clearly distinguishes Lecture Notes from journal articles or technical reports which normally are very concise. Articles intended for a journal but too long to be accepted by most journals, usually do not have this "lecture notes" character. For similar reasons it is unusual for doctoral theses to be accepted for the Lecture Notes series, though habilitation theses may be appropriate.

2. Besides monographs, multi-author manuscripts resulting from SUMMER SCHOOLS or similar INTENSIVE COURSES are welcome, provided their objective was held to present an active mathematical topic to an audience at the beginning or intermediate graduate level (a list of participants should be provided).

   The resulting manuscript should not be just a collection of course notes, but should require advance planning and coordination among the main lecturers. The subject matter should dictate the structure of the book. This structure should be motivated and explained in a scientific introduction, and the notation, references, index and formulation of results should be, if possible, unified by the editors. Each contribution should have an abstract and an introduction referring to the other contributions. In other words, more preparatory work must go into a multi-authored volume than simply assembling a disparate collection of papers, communicated at the event.

3. Manuscripts should be submitted either online at www.editorialmanager.com/lnm to Springer's mathematics editorial in Heidelberg, or electronically to one of the series editors. Authors should be aware that incomplete or insufficiently close-to-final manuscripts almost always result in longer refereeing times and nevertheless unclear referees' recommendations, making further refereeing of a final draft necessary. The strict minimum amount of material that will be considered should include a detailed outline describing the planned contents of each chapter, a bibliography and several sample chapters. Parallel submission of a manuscript to another publisher while under consideration for LNM is not acceptable and can lead to rejection.

4. In general, **monographs** will be sent out to at least 2 external referees for evaluation.

   A final decision to publish can be made only on the basis of the complete manuscript, however a refereeing process leading to a preliminary decision can be based on a pre-final or incomplete manuscript.

   Volume Editors of **multi-author works** are expected to arrange for the refereeing, to the usual scientific standards, of the individual contributions. If the resulting reports can be forwarded to the LNM Editorial Board, this is very helpful. If no reports are forwarded or if other questions remain unclear in respect of homogeneity etc, the series editors may wish to consult external referees for an overall evaluation of the volume.

5. Manuscripts should in general be submitted in English. Final manuscripts should contain at least 100 pages of mathematical text and should always include

   - a table of contents;
   - an informative introduction, with adequate motivation and perhaps some historical remarks: it should be accessible to a reader not intimately familiar with the topic treated;
   - a subject index: as a rule this is genuinely helpful for the reader.
   - For evaluation purposes, manuscripts should be submitted as pdf files.

6. Careful preparation of the manuscripts will help keep production time short besides ensuring satisfactory appearance of the finished book in print and online. After acceptance of the manuscript authors will be asked to prepare the final LaTeX source files (see LaTeX templates online: https://www.springer.com/gb/authors-editors/book-authors-editors/manuscriptpreparation/5636) plus the corresponding pdf- or zipped ps-file. The LaTeX source files are essential for producing the full-text online version of the book, see http://link.springer.com/bookseries/304 for the existing online volumes of LNM). The technical production of a Lecture Notes volume takes approximately 12 weeks. Additional instructions, if necessary, are available on request from lnm@springer.com.

7. Authors receive a total of 30 free copies of their volume and free access to their book on SpringerLink, but no royalties. They are entitled to a discount of 33.3 % on the price of Springer books purchased for their personal use, if ordering directly from Springer.

8. Commitment to publish is made by a *Publishing Agreement*; contributing authors of multiauthor books are requested to sign a *Consent to Publish form*. Springer-Verlag registers the copyright for each volume. Authors are free to reuse material contained in their LNM volumes in later publications: a brief written (or e-mail) request for formal permission is sufficient.

Addresses:

Professor Jean-Michel Morel, CMLA, École Normale Supérieure de Cachan, France
E-mail: moreljeanmichel@gmail.com

Professor Bernard Teissier, Equipe Géométrie et Dynamique,
Institut de Mathématiques de Jussieu – Paris Rive Gauche, Paris, France
E-mail: bernard.teissier@imj-prg.fr

Springer: Ute McCrory, Mathematics, Heidelberg, Germany,
E-mail: lnm@springer.com

MIX
Papier aus verantwortungsvollen Quellen
Paper from responsible sources
FSC® C105338

If you have any concerns about our products,
you can contact us on
ProductSafety@springernature.com

In case Publisher is established outside the EU,
the EU authorized representative is:
Springer Nature Customer Service Center GmbH
Europaplatz 3, 69115 Heidelberg, Germany

Printed by Libri Plureos GmbH
in Hamburg, Germany